A-Z
ZOOLOGY

A-Z
ZOOLOGY

Prof. Nirmal Chandra Pradhan

CENTRUM PRESS
NEW DELHI-110002 (INDIA)

CENTRUM PRESS
H.O.: 4360/4, Ansari Road, Daryaganj,
New Delhi-110 002 (India)
Ph.: 23278000, 23261597

B.O.: No. 1015, Ist Main Road, BSK IIIrd Stage
IIIrd Phase, IIIrd Block,
Bangalore - 560 085 (India)
Tel.: 080-41723429
Visit us at: www.centrumpress.com

A-Z Zoology

First Edition, 2009

ISBN 978-93-80106-41-0

PRINTED IN INDIA

Printed at Salasar Imaging Systems, Delhi-110035 (India)

Contents

Preface

Biology or Life science has made rapid progress. Advancements in several fields of Science and Technology have provided the tools to understand life processes. Zoology concerned with the origin of species from a common descent, and descent of species; as well as their change, multiplication, and diversity over time. Scientific and metaphysical theories on origin and the evolution of creatures lead to the anatomical and fossil evidences on the evolution from worms to vertebrates with backbone and brains by way of cells, tissues, and organic correlation and comparison of skeleton and head. Development of mind and the sequence of mental functions and natural selection and the role of environment in evolution of species is discussed afterwards.

Anatomical figures, wherever possible, were made from special dissections and those of bilateral animals are shown from the left side to facilitate comparisons. In series showing embryological or larval development the individual figures are brought into comparable positions so that changes during growth or metamorphosis can be traced or compared directly with drawings of the adult animals. In series showing embryological or larval development the individual figures are brought into comparable positions so that changes during growth or metamorphosis can be traced or compared directly with drawings of the adult animals.

Author

Chapter 1

Introduction to Zoology

Zoology or Animal Biology is that branch of science which deals with animals. About half a million different sorts or species of animals have already been described and named, each breeding true to its own special characteristics, each different from all others. The number of fresh species discovered, described, and named every year is about two hundred, and this number is at present increasing, not decreasing, year by year.

Animals inhabit sea, fresh water, land, and air, or they may live on or in the bodies of other animals or plants. Their range of size is enormous. The malarial parasite is so small as easily to inhabit the interior of a human red bloodcorpuscle, of which five million are normally contained in a cubic millimetre of blood; while the sulphur-bottom whale (*Balaenoptera sulphureus*), the largest animal known, may reach a length of 95 feet, and a weight of 147 tons, or nearly three times as much as most express engines. Their shape is as various as their habits or their size.

Some, like certain radiolaria, form beautiful geometrical designs; others are almost shapeless, like many parasites and sponges; some resemble plants; still others, such as the nematode worms, look like long threads; let alone all those innumerable shapes of bird and fish and mammal, crab and spider and insect, that we all know.

It is obvious that any study of all these creatures, their structure and mode of working, their habits and their history, will soon give us an enormous body of facts which will be overwhelming unless we classify them properly. Broadly speaking, we want, first of all, to find out how a particular animal works, considered as a piece of living mechanism; and to compare the ways of working of various animals. That is Animal Physiology. And secondly, we want to know all we can about the structural plan of animals, to know how that structure develops, and to compare the structure of different animals. That is Animal Morphology, the science of form. Finally, we want to understand, if possible, how and why it is that the different individuals and species of animals are what they are—their history and as much as possible of the causes of that history. That is the science of Animal Evolution and Heredity, sometimes called Genetics (although this term is often restricted to Heredity alone).

There remains the problem as to the fundamental difference between plants and animals. Why do we call this organism a plant, this other organism an animal? Most people would not hesitate, but would say that the animal moved while the plant did not; that the animal was conscious while the plant was not; that the animal devoured its food while the plant absorbed its nutriment from its surroundings.

None of these criteria, however, is absolute. Many animals, like coral-polyps or sea-squirts, are as rooted to the spot as most plants; while some undoubted plants move about. He would be a very bold man who asserted that a sponge, an undoubted animal, possessed a higher level of consciousness than a mushroom or a wall-flower; while many animal parasites absorb their food from the medium which bathes them.

As a matter of fact, the only valid distinction between plants and animals is concerned with the type of foodstuffs which they can utilize. All organisms, plant or animal, need carbon, hydrogen, nitrogen, and oxygen to build the bulk of their bodies. Green plants can, with the aid of sunlight, obtain carbon from the carbon dioxide of the air or water in which they live. They can obtain their hydrogen from water and salts, their oxygen directly from the air, their nitrogen from simple mineral salts like nitrates. In other words, the green plant can build up living protoplasm from elements and the simplest compounds. Every particle of living matter added to a green plant means the creation of a new kind of material combination. Animals on the other hand cannot achieve this synthesis.

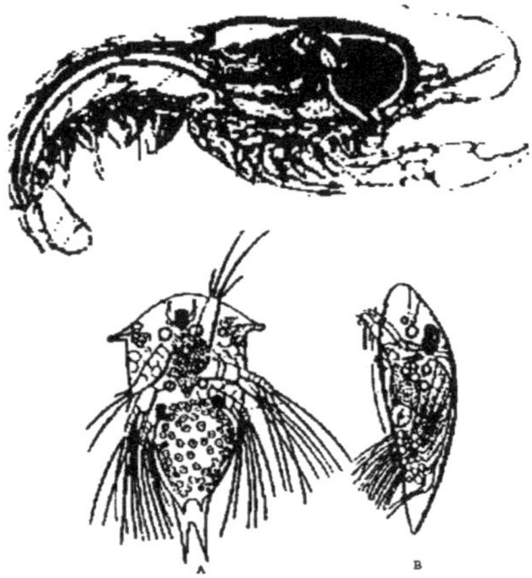

Fig. Parasitism and Recapitulation

Earliest free swimming stage or Nauplius with three pairs of appendages. In this stage it closely resembles the larvae of many other Crustacea. Later stage, in which it attaches itself to its host.

They have to be provided with highly elaborate compounds, all of which in the long run owe their existence to the manufacturing powers of plants. An animal cannot obtain its carbon from any compound less complex than a sugar, a starch, or a fat; for its nitrogen it must be provided with proteins, or at least with the constituent parts of proteins known as amino-acids, which are already of considerable complexity. Animals are in the long run always dependent upon green plants; they are, one might almost say, parasitic upon plants. Green plants by the same token are parasitic upon the sun; they live by stealing energy from his rays.

There are other plants besides green plants: fungi and bacteria contain no chlorophyll. Most fungi are as dependent as are animals upon the previous activities of other organisms; they can only live where decay has provided them with raw materials. But even so they are not so helpless as animals, and can obtain their food from less complex compounds. Among the bacteria are forms which show quite extraordinary modes of nutrition.

The most interesting are those which can directly fix and utilize the nitrogen of the air, a process for whose accomplishment man has to apply enormous stores of energy. Others can utilize carbon dioxide as their source of carbon without making use of chlorophyll. Still others can live without free oxygen, and obtain all they want from chemical compounds. However, the great bulk of the food-cycle of the world starts with the activities of green plants, and all animals, all fungi, and most bacteria are in a very real sense dependent upon the green plants' chlorophyll.

This is the basic distinction between animals and plants, and all the differences with which we are so familiar, between higher plants and higher animals, are purely secondary. The fact that green plants can obtain food from water and air, without special search, has led to their developing great feeding surfaces—such as the leaves and the roots—in air and water respectively. The fact that animals have to find their food ready-made has led to their developing mouths and stomachs to catch and hold the food, and limbs to move from place to place in search of more. The fact of locomotion has in its turn made necessary the development of senseorgans and nervous system and brain. But all hinges on the first and most vital difference.

Even so, there are some types, among the simplest and smallest creatures, which share animal and plant characteristics, being able both to take in solid food like a typical animal, and to build up food from simple inorganic substances like a green plant. Such examples only show the impossibility of drawing hard and fast lines in Nature. But these are all generalities, and in order to be able to deal properly with what is general, we must have a good acquaintance with the particular.

The best way of doing this will be to take a single species of animal and describe its structure and working in broad outline. For our purpose any one of the higher animals would really serve, but on account of the ease with which it can be obtained, its convenient size, and the resemblance of its general structure and functions to those of man, we will take the frog as our introductory.

THE GENERAL ANATOMY AND PHYSIOLOGY OF THE FROG

At first reading, the statement that frogs resemble men in any important degree may perhaps raise a smile. It is nevertheless true. We can recognize in the frog a great many parts that exist in ourselves, arranged moreover in the same way. A frog possesses a head, a trunk, and fore and hind limbs. The nostrils, eyes, ear-drums, and mouth are arranged in the same relative positions as in our head.

If we look at the skeletons of man and frog, we shall find that both possess a skull, a backbone consisting of separate jointed pieces or vertebrae, the same type of limb-bones, the same kind of teeth. If we dissect them, we shall in both discover red blood, a heart in the front of the trunk and on the ventral surface, a liver, a pair of kidneys, a spleen, a nerve-cord within the backbone, and a great many other organs which have a family likeness to each other, and are to be found in similar positions in the bodies of the two organisms. The same plan is also found in the horse.

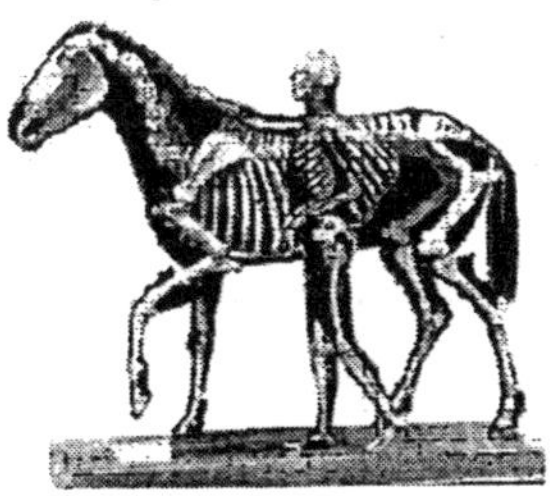

Fig. Skeletons of Man And Horse

But if we had chosen a crayfish, say, as our type, these correspondences would not have been there. A crayfish possesses not two but nineteen pairs of limbs. It has no backbone, but grows its skeleton on the outside.

It has a heart, but it is in the centre of the body, and towards the back or dorsal side; its blood is nearly colourless; it has no nostrils or ear-drums; no spleen; the nerve-cord runs down its ventral side instead of along its back; its kidneys are in its head—in fact, it is difficult to find any point in which its plan of structure closely resembles that of man.

Fig. Diagrammatic Dissection of a Female Crayfish

We shall come back to this question of the resemblances and differences between animals. Now we must return to the frog, and ask ourselves what it does and how it does it. Like other animals, the frog eats; it breathes; it must get rid of waste; it must move in order to procure its food or to escape its enemies or to find its mate; changes in the outer world affect it; there must be some means by which the parts of its body can be made to act together as a whole, instead of merely as a number of separate parts; and finally, it reproduces its kind.

Why does the frog, or indeed any other animal, require food? It requires it for two main reasons. First, the frog is doing physical work every time it moves; to do work it needs some source of energy; and, as a matter of fact, it obtains this energy by the oxidation or slow combustion of some of the substances contained in its food. Secondly, the substance out of which it is made is all the time slowly wearing out or breaking down, and needs to be repaired continually by other substances out of the food.

The living machine thus burns part of its food for fuel, and uses other parts for repairs. The frog feeds on worms, small snails and slugs, insects, and other small animals. It seizes them with its tongue, which is sticky and attached at the front instead of at the back, and can be suddenly shot out of the mouth. Once in the mouth, the prey is held not only by the teeth, which are small, all alike in shape, and only to be found on the upper side of the mouth, but also by the eyeballs, which, unlike our own, can be brought right down into the cavity of the mouth.

At the back of the mouth the prey is forced into the opening of a narrow tube, the gullet, which leads down into the saclike stomach.

Once any solid object is inside the gullet, this contracts automatically in a series of waves, driving the object downwards and into the stomach. Out of the far end of the stomach opens a coiled narrow tube, the small intestine; but the opening can be closed by a ring of muscle called a sphincter, and as a matter of fact the prey is kept in the closed stomach for some time.

During this time it is exposed to the action of a juice, the gastric juice, which is manufactured by the walls of the stomach, and as a result. It becomes largely dissolved. When it is reduced to a pulpy broth, it is passed on to the intestine.

Fig. General Anatomy and Arteries of a Male Frog

Into the beginning of the intestine there open two very small tubes or ducts. One is the bile-duct, which leads from the liver, a very large brownish organ divided into several lobes; the bile, which is produced by the liver, is a green fluid, and is stored until wanted in a round vessel, the gallbladder, connected with the bile-duct. The other is the pancreatic duct, which comes from a small pinkish-white organ, the pancreas.

The bile and pancreatic juice, together with a juice derived from the intestine itself, complete the work begun in the stomach, until finally all of the food that is available for the use of the body is dissolved. It can now be passed through the living wall of the intestine into the blood and so distributed to the rest of the body.

The process of rendering the food soluble is what we call *digestion*. This is completed in the first part of the small intestine, while *absorption* takes place in the remaining parts. When all the absorption that is possible has taken place, there still remains some residue, indigestible and useless to the animal. This is called the faeces; it passes from the small intestine into the broader and shorter large intestine or rectum; here it is consolidated into pellets, and is eventually passed out between the frog's legs at the opening of the cloaca, from the Latin word for a sewer.

There is thus a tube, the digestive tube, running from mouth to cloaca. Its cavity is open to the exterior at both ends, and so is, in a certain sense, not inside the frog at all. Digestion is simply the process of turning the food into a condition when it can be passed, by absorption, into the real interior of the body.

It would be perfectly possible for an organism to absorb food over the whole surface of its body, and some animals do so. But in a creature like the frog it is obviously important that the part of it which is directly exposed to the outer world should act as a protective covering. Accordingly we have the external surface covered by the protecting skin, while the duties of digestion and absorption, which demand more delicate tissues, are carried out by an internal surface, the lining of the gut.

The stomach and intestine lie in a space, the general bodycavity or coelom; they are kept in place within it by a delicate fold of membrane, the mesentery, which connects them with the dorsal side of the body-cavity. In this membrane may be seen a large number of red tubes—blood-vessels conveying the red stream of blood, always in the same direction in any one tube.

On the gut itself a meshwork of very small vessels as a matter of fact their smallest microscopic branches, or capillaries as they are called, come into close connexion with the lining of the digestive tube, and the dissolved food-substances are passed through their walls into the blood. The small branches can be seen to unite into larger, and these into larger still, until finally the whole of the blood from the gut is seen to pass into the liver by a single vessel.

Before we pursue the fate of the blood in the liver, we must consider the general plan and working of the bloodsystem. The web of a frog's foot or the tail of a tadpole is transparent enough for us to see its blood-vessels under the microscope.

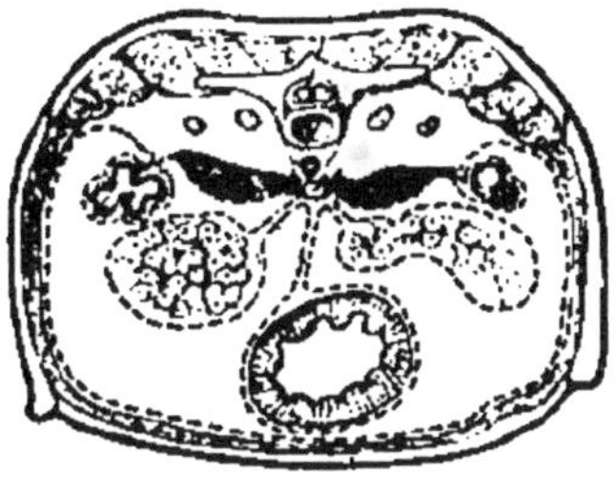

Fig. Section Across the Hinder end of The Trunk in A Female Frog

We see solid particles, the blood-corpuscles, hurried along within these various-sized tubes in a stream whose motion is always in one direction, and takes place by jerks. The blood-vessels are branched, and in some of them the blood passes from large trunks to smaller and smaller branches, in others from the small branches to the main trunks. The former sort of vessel is generally called an artery, the latter a vein. If a frog is dissected, most of the large trunks are found to end in the heart; if this is opened, it is found to be nothing more nor less than a hollow bag of muscle, divided into several chambers.

It is so constructed that when the muscles contract, or in other words the heart beats, the blood is driven through it, always in the same direction, owing to the arrangement of valves within it. From what we said above, it is clear that the blood leaving the heart will pass into the main arteries, and that blood will be pushed in from the main veins to take its place. The blood moves in jerks because of the successive beats of the heart, and the net result of the working of the system is that blood is continually circulating from the capillaries to the heart and back again. This simple fact of the blood's circulation, although at the bottom of any real knowledge of physiology, was not discovered until the early seventeenth century, by William Harvey.

There are three pairs of arteries leaving the frog's heart. One divides into branches supplying the mouth, head, and brain. The next pair is the largest; the two members of the pair unite to a common trunk running along the back, and called the dorsal aorta. This pair sends branches to both limbs, to the digestive system and all other internal organs, and to the muscles of the body. The third pair sends one branch to the lungs and another to the skin. The three pairs between them supply blood from the heart directly or indirectly to every organ of the body.

In the organs, the smallest branches of the arteries divide into capillaries and the blood is driven on from these into the small veins. The system of veins is more complicated than that of the arteries. The blood from the head, fore-limbs, skin, and lungs passes directly to the heart; but that from other parts travels a more complicated route.

The blood from the capillaries of the digestive system, as we saw, passes into the liver in a large vein. In the liver, this branches and forms capillaries again, new veins are once more formed from these, and the blood only reaches the heart after having passed through two sets of capillaries instead of one. Such a vein is called *portal*; and we have thus the portal vein of the liver. There is also in the frog (but not in man) a portal vein of the kidneys, which leads most of the blood from the capillaries of the hind-limbs to a second set of capillaries in the kidneys.

The living tissues of every part of the body are thus in contact with capillary blood-vessels, and these have such thin walls that soluble substances can diffuse through them from the slow-moving blood in them to the tissues or from the tissues to the blood. Since the capillaries are all part of the single blood-system, and the blood is always in circulation, it follows that substances from any part of the body can be transported to any other part.

The blood-system is thus, among other things, the body's system of distribution and exchange. It plays roughly the same part in the body of a frog or a man as is played in a modern nation by the traffic of railways, roads, and canals, the markets and retail tradesmen, the last-named being represented by the capillaries. What is it that the blood distributes? In the first place, food.

The dissolved food from the gut is taken to the liver; this acts as a sort of warehouse and refinery. Some surplus food is stored there to be distributed gradually as needed, and other food substances are chemically changed by its action. The next substance to be distributed is oxygen.

It is obvious that energy is needed for carrying out movements, and as a matter of fact it is provided by the combination of oxygen with substances in the muscles; the more muscular work is done, the more oxygen is needed; the brain, too, is very sensitive to lack of oxygen; indeed we can say that the general processes of life in higher animals are only possible as a result of steady oxidation.

For it to pass into the blood there is needed a moist membrane, very thin, with oxygen on the one side and capillaries on the other. Since oxygen exists in the air, it would be possible for the skin to be utilized as such a membrane; and this does actually happen in the frog.

Its skin is very richly supplied with blood-vessels, and is always moist. As a result, frogs can only live in damp places. Higher organisms, such as ourselves, have a stronger skin, and one which is dry. They can, therefore, live in more varied surroundings, but can no longer use their skin for absorbing oxygen.

The frog, however, does not rely entirely on its skin, and it also possesses lungs, which can best be thought of as an internal surface specially designed for exchange of gases between blood and air. The lungs in both frog and man are a pair of spongy thin-walled bags divided up into a great number of compartments (and so providing a great deal of surface) in whose walls run very many blood-vessels. They are put into communication with the air by means of a tube, the wind-pipe or trachea, which opens into the back of the mouth-cavity just in front of the gullet.

In ourselves, air is sucked into and forced out of the lungs mainly by the movements of the chest and diaphragm, to whose walls the lungs are attached. But in the frog, air is sucked into the mouth through the nostrils, and then forced down into the lungs by contraction of the muscles of the throat, the nostrils being at the same time closed; it is driven out again by the elasticity of the lungs themselves. The frog has no diaphragm. The difference between our method and the frog's is like that between a suction-pump and a force-pump.

The oxygen passes into the blood-system through lungs and skin, and, like the food, is distributed by the circulating blood to all parts of the body. Here it enters into combination with various constituents of the living substance, and, as a final result of these chemical processes, waste-products are produced, which damage the organism if they accumulate, and must be got rid of.

The most important of these endproducts of life's activity are carbon dioxide (CO_2), water (H_2O), and urea (N_2H_4 CO); and the process of ridding the body of such substances is called *excretion*. Carbon dioxide is a gas, and its excretion can and does take place through the same membranes, of lungs and skin, which serve for the intake of oxygen.

Urea, however, and any surplus salts are not gaseous and so can best be excreted in solution. The chief organs which remove substances from the blood in solution are the kidneys. In the frog, these are found at the back of the coelom. They consist of a great number of microscopic tubes, twisted together, and richly supplied with blood.

The tubes eventually all open into a large draining-tube, or duct, which runs backward and opens into the cloaca. Surplus urea and salts, together with water, are taken up from the blood by the little tubes, and the resulting fluid or urine is drained out along the duct.

Just opposite the openings of the ducts into the cloaca is another opening, that of the bladder, which is thin-walled and muscular, and lies on the front of the large intestine. This is simply used to store the urine until a considerable amount has accumulated, when it is passed out of the cloaca.

Finally, not all the surplus water is excreted by the kidneys; some is got rid of in the form of water-vapour by lungs and skin. The whole of the chemical processes going on in an organism are known collectively as its metabolism. This consists partly of the building-up of the soluble food-materials into very complex colloid molecules, of which the living framework consists, partly in the breakdown and wastage of this framework, partly in the breakdown of the simpler substances which act as fuel for energy-production.

If we now turn back for a moment to the blood-system, and consider its detailed arrangement. It will be easier to this from the blood-system of man, which is in some ways both simpler and more efficient. Here the heart consists of two separate halves, a right and a left, each consisting of a thin-walled chamber or auricle opening into a thicker-walled ventricle. The veins enter the auricles, the arteries leave the ventricles; and there exist flaps of membrane which act as valves and only allow the blood to pass in the one direction.

The only veins which enter the left auricle come from the lungs; they therefore contain blood rich in oxygen and poor in carbon dioxide. In this condition blood is called arterial. From the left auricle it passes on into the left ventricle, and thence into the main artery or aorta, whose branches carry blood to all the organs with the single exception of the lungs.

In the capillaries of the organs, the living tissues take the oxygen they need from the blood, and discharge into it the carbon dioxide they have produced. The resultant blood, poor in oxygen but rich in carbon dioxide and other waste-products, and called venous blood, is collected in the veins, and is sucked into the right auricle. Thence it is pumped into the right ventricle, and so through an artery to the lungs.

The way in which the system of pump and tubes is constructed thus ensures that all blood which has given away oxygen to the organs of the body shall go to the lungs, to be charged again with oxygen and to be rid of carbon dioxide, before going out once more to any of the other organs.

The portal vein takes all the blood from the digestive system to the liver so that the surplus food materials may be there dealt with at once before going to the rest of the tissues; the liver is thus in one respect like a central storehouse from which certain food-stuffs are rationed to the rest of the body as required.

In the frog the plan of the blood-system is slightly different. Both auricles open into one single ventricle, so that some mixing of venous and arterial blood takes place. From the ventricle springs a tubular part of the heart, or *truncus*, not found in man.

The position of the aperture from ventricle to truncus, and the valves inside the truncus, are so arranged, however, that the most arterial blood passes into the artery leading to the head, the mixed blood into that supplying the limbs and body, the most venous blood into that leading to the lungs. Thus the brain gets the blood richest in oxygen, and most, but not all, of the venous blood is taken to the lungs before again going to body or head.

The blood-system, however, is not only concerned with transport. Organisms are faced with the problem of coordination: and the blood-system provides one method of dealing with this. The problem is this: given a number of organs, such as heart, lungs, limbs, stomach, kidneys, brain-how to ensure that they shall work together for the good of the organism, and not simply pursue their own activities independently of each other—how, in other words, to convert a mob into an army.

The way in which the pancreas is made to secrete its digestive juice at the right time, and only at the right time, will provide us with a good example of co-ordination through the blood-stream. The pancreas is usually inactive; but the passage of food from the stomach into the intestine is known to be followed by a secretion of pancreatic juice which is poured down the duct to help digest the food.

How is this done? When the food passes into the intestine, it stimulates the intestine chemically, causing it to secrete a special substance from its lining; this passes into the blood, circulates through the whole body, but, though it exerts no effect on most organs, stimulates the pancreas (and probably the liver) to activity.

This substance is called 'secretin'. It can be artificially

extracted from the lining of the intestine and will then, if injected, cause the pancreas to secrete. Such 'chemical messengers' are called *hormones*, and the blood provides the channel by which they exert their chemical co-ordination between parts of the body. As they are secreted into the blood, and not down a duct on to some free surface, they are included under the term *internal secretions*.

Other internal secretions regulate growth and metabolism, prevent one organ from growing disproportionately to the rest, or have a say in the rate at which the various chemical processes of life shall work. A substance secreted by the pituitary, for instance, which is a small gland at the base of the brain, influences the growth of bone.

If too much of it is present in youth the bones grow excessively, and giants are the result. Another substance secreted by the same gland causes frogs to become darker in colour. The thyroid gland is situated in the neck region of Vertebrates.

Its secretion influences the rate at which their metabolism goes on; with most mammals, too much makes them nervous and excitable, too little leaves them sluggish in mind and body. The adrenals, the parathyroids, the pancreas, the reproductive organs, and probably other organs also produce internal secretions.

Finally, the blood helps in the defence of the body. If proteins which are not normally found in a certain organism are injected into it, they are precipitated or broken down into simpler substances. Not only that, but if they are injected a second time after a proper interval, they can be destroyed more rapidly and in greater quantity.

Bacteria, many of which, if they could live in the tissues or the blood, would give rise to diseases, contain such foreign proteins; and in Nature it is chiefly bacteria which are thus destroyed if they obtain an entrance into the tissues. Familiar examples of the utilization of this property are vaccination, preventive inoculation for typhoid, and the antitoxin treatment of diphtheria.

The blood is thus the great distributor; it distributes the

raw materials of life, its waste products, the chemical substances concerned in co-ordination and regulation, and those which help in the protection against disease. It is a middleman between each living part of the body and every other, and its circulation, begun in the first weeks of life, must continue uninterruptedly if life is to be maintained. The only rest which the heart can have is between each beat and the next.

There is in the frog, as in man, another set of spaces filled with fluid which are circulatory in function. These are the lymphatics. Into them any surplus fluid which has passed from the blood into the tissues is drained out, and this fluid is not red but colourless.

In man the lymphatics start as small irregular spaces, which unite and eventually drain into one of the large veins. In the frog, however, the small lymph-spaces unite into very large lymph-sacs, the biggest of which, filled with clear fluid lymph, lie between the skin and the muscles of the body-wall.

To pass lymph from these into the veins, special lymph-hearts exist—two pairs of small muscular sacs pumping after the fashion of the true heart. The faint pulsation of one pair of these can be seen in life, just anterior to the cloaca, on the dorsal surface between the backbone and the hip-bones: the other pair is below the shoulder-blades.

The frog can move from place to place; and special organs are needed for movement as for digestion and circulation. The actual parts of the frog by which movement is effected are the muscles or flesh. Muscle is a form of living substance which has the property of contraction, or altering its shape, when stimulated in certain ways; it shortens its length, while increasing in breadth.

There is, for instance, a layer of muscle arranged circularly round the gullet (as also round the rest of the gut). When this contracts in any one place, it narrows the tube of the gullet there; and it is by a wave of such contraction travelling from top to bottom of the gullet-muscles that food is automatically passed down into the stomach. Or again, when the muscles of the bladder contract, the cavity of the bladder is made smaller and urine is expelled. For moving the animal from place to place on land, however, some part of the body must be held fixed against the ground, and the rest of the body moved relative to this fixed point. To accomplish this, a jointed framework is necessary; and this is provided in the frog by the skeleton, composed of substances known as bone and cartilage.

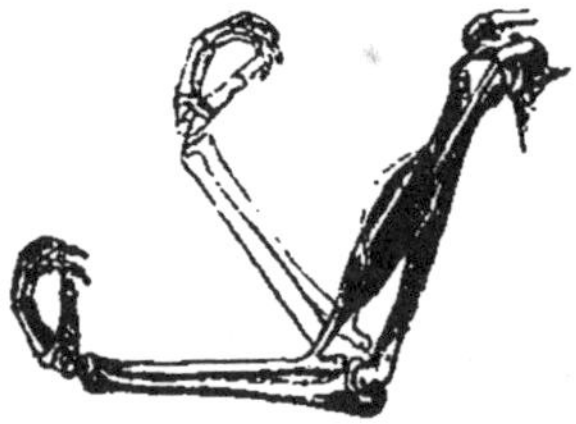

Fig. The Biceps and Skeleton of the Human Fore-limb

The skeleton, however, serves other purposes besides enabling the contraction of the muscles to effect movement. For one thing it acts as a support to the whole body.

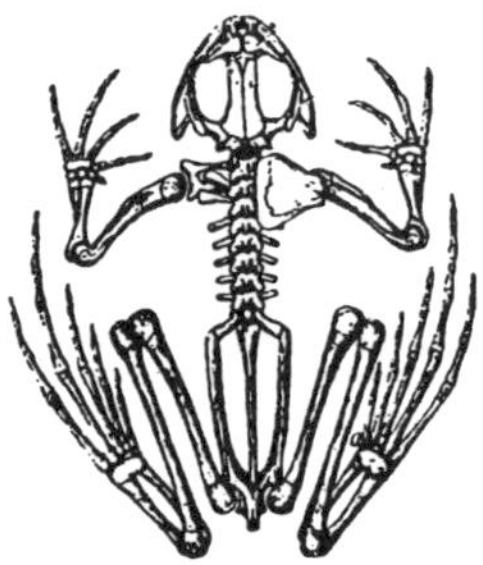

Fig. The Skeleton of a Frog.

Living substance itself is soft and semi-fluid, with a specific gravity very slightly greater than that of water. Animals which live in the water, therefore, are almost entirely supported by the water; but a land animal of any size requires a firm skeleton to prevent it collapsing under the force of its own weight.

A frog without a skeleton would spread out, if in air, like an egg taken out of its shell; whereas jelly-fish far bigger than frogs can manage to preserve their shape in water with no support except their watery jelly.

Again, certain delicate organs require to be protected against injury and shock; in the frog, the brain and spinal cord, the organs of smell and hearing, and to a certain extent the eye, are enclosed within parts of the skeleton.

The frog's skeleton is built on the same general plan as that of a man. There is first a central portion consisting of skull and backbone. The skull is a composite structure. In the centre is the brain-box or cranium, containing the whole of the brain. In front of these are the two nasal capsules, closed above and at the sides, open below, in which the organs of smell are lodged. At either side of the cranium's hinder end are similar but still more completely closed capsules, to house the organs of hearing.

The skeleton of the upper jaw is fixed to the nasal capsules in front and to the ear-capsules behind. On it are numerous small teeth, all of the same pattern, and merely attached to the surface of the jaw, instead of being firmly fixed in sockets like our own. There are teeth also on a pair of small bones near the centre of the roof of the mouth, but none on the lower jaw-bone, which is hinged to the hind end of the upper jaw. The orbits, or spaces for the eyeballs, lie between the cranium and the upper jaw. There is finally the hyoid, a small plate embedded in the floor of the mouth and attached to the hinder end of the skull; this gives attachment to the muscles which move the tongue.

The backbone encloses the spinal cord. It is formed of nine separate pieces or vertebrae, together with a longer rod at the hind end which represents a number of vertebrae joined together. Each vertebra consists essentially of a more solid ventral piece, mainly for support, to whose upper side is attached an arch for the protection of the spinal cord. The first vertebra fits on to the hind end of the skull, and the others are jointed to each other by ball-and-socket joints in the supporting portions, and by pairs of smooth surfaces springing from the arches and fitting one against the other. There are no ribs in the frog.

The skeleton of the hind-limb consists of a thigh-bone, fitting at its upper end into the hip-girdle; a shank-bone; the ankle-bones; and the bones of the foot, each toe containing a series of small bones arranged in a row. All these parts are jointed to each other in various ways. The joint between thigh and hip-girdle is arranged on a ball-andsocket principle, allowing the thigh to be moved into a great variety of positions; those between thigh and shank or shank and ankle, however, while allowing free motion backwards and forwards, do not allow much in other directions. A number of muscles are to be found in the hind-limb. They may be attached directly to a part of the skeleton, or indirectly by means of a tendon, a tough sinewy cord, which may be fixed to a bone some distance away, and so act like a pulley-rope.

The skeleton of the fore-limb is built on exactly the same plan, but all its parts are shorter, particularly that part (the wrist) which corresponds with the ankle in the hind-leg. All animals which jump have their jumping legs elongated in order to give powerful leverage; one has but to think of a grasshopper, a kangaroo, or a jerboa, besides the frog.

The hind-legs are attached to the hip-girdle, and this in its turn is firmly attached to part of the backbone. The forelimbs' skeleton is also attached to another portion of the skeleton, the shoulder-girdle; this, however, is not attached to the backbone, but is simply embedded in the muscles of the body-wall. The shoulder-girdle also joins ventrally on to the breast-bone or sternum, which protects the heart, and gives attachment to muscles moving the fore-limbs and the hyoid.

By means of the muscles and the skeleton, then, the frog can move. But how are its movements to be regulated, made to serve some useful purpose such as capturing food? How are the muscles to be co-ordinated together instead of one working independently of, or even in antagonism to, another? This is effected by means of the nervous system and the receptor organs.

The receptor organs are those parts of the living organism which are specially sensitive to the changes going on around them. Some of them are affected by the changes going on inside the body in muscles and joints and in the organ of balance (proprioceptors), others by the changes taking place in the world outside (exteroceptors). Some of these latter are especially sensitive to changes of temperature, others to changes of pressure; some to waves in the ether (light), others to waves in the air; others again to chemical substances; some translate all changes which affect them into a sense of pain.

The exteroceptor organs are the windows of the animal into the outer world. Through the proprioceptors we are aware of the position of the various parts of the body, which of course depend on the degree of contraction of a number of muscles, the degree of bending of the various joints, and upon our position with regard to the vertical. The receptor organs include the sense organs, but are not the same thing, since many receptors when stimulated do not always give rise to sensations, and some never do.

A receptor organ, in fact, is in itself responsive to one particular sort of change, but its stimulation may or may not give rise to a sensation in consciousness. They enable action to take place in response to changes inside or outside the body, and in some cases in addition the animal is through them made aware of these changes, by sensations being aroused. Receptor organs may be large and important structures; such, in the frog or man, are the eye, the ear, and the organ of smell. By means of the eye it is possible for the frog to be aware of the form, size, and probably colour of objects at a distance; if it had no ear it would not only be unable to perceive sounds, but also to balance itself.

The nose enables it to detect distant objects by reacting chemically with particles which they give off into the air. Taste, like smell, is a chemical sense, but gives information not about distant objects but about those which find their way into the mouth; a number of very small taste-organs are scattered over certain parts of the tongue. The receptor organs for touch, pain, heat, and cold are all microscopic, and are scattered over the surface of the body, more abundantly in some regions than in others.

In all the higher animals, receptor organs are always connected with the nervous system. This consists of the central nervous system, and the nerves and ganglia. The central nervous system includes the brain and the spinal cord. The brain of a frog is a soft whitish organ, richly supplied with blood, and of a complicated shape. We can distinguish three main divisions in it, the fore-brain, the mid-brain, and the hind-brain. In the fore-brain, the largest part is the cerebrum, consisting of the paired cerebral hemispheres. In front of this are olfactory lobes connected with the nerves of smell; behind it a small part with curious stalked bodies arising from it, the pineal above and the glandular pituitary below. The mid-brain is small; the hind-brain again large, and divided into the cerebellum and the medulla.

The spinal cord is joined to the medulla. It runs the length of the backbone, and has no specially distinguishable parts. From both brain and spinal cord spring a number of white branching structures, the nerves. Those from the brain are called the cranial nerves. When traced out, most of them are found to end in the sense-organs and muscles of the head, but one pair in particular, the vagus nerves, run right down into the body and send branches to heart, lungs, stomach, and other organs.

On the other hand, none of the spinal nerves, from the spinal cord, run up into the head (except the first, which supplies the throat region); they end in the receptor organs of the body and in the muscles of the limbs and body-wall.

Down the back of the body-cavity there may also be seen a double chain of nerves which is not directly connected with either brain or spinal cord. On this chain there is typically a swelling or ganglion opposite each spinal nerve, and this ganglion is connected with the corresponding spinal nerve by a thin nerve-branch; the chain also continues into the head, where similar connexions are made with some of the cranial nerves. This set of nerves is called the sympathetic system; the branches given off from it run mainly to glands and to muscles not connected with the skeleton but forming part of internal organs such as those of bloodvessels, of the digestive tube and of the bladder.

In the brain the cerebral hemispheres (*He*) are shown, with the olfactory lobes in front of them. Behind the X-shaped figure made by the junction of the optic nerve lies the pituitary. *M* marks the junction of the medulla oblongata and spinal cord.

What is the function of the nervous system? When we examine a nerve we find that it is composed of a bundle of nerve-fibres, in the same way as an electric cable is composed of a bundle of wires. Each of these nerve-fibres is a microscopic thread of living substance which has the power of conducting impulses or excitations very rapidly along its length, roughly as a wire conducts an electric current. And the nerves themselves are like cables each containing a number of wires. The nerve-fibres from receptor organs run in nerves until they reach the central nervous system, and there split up into a number of very fine branches.

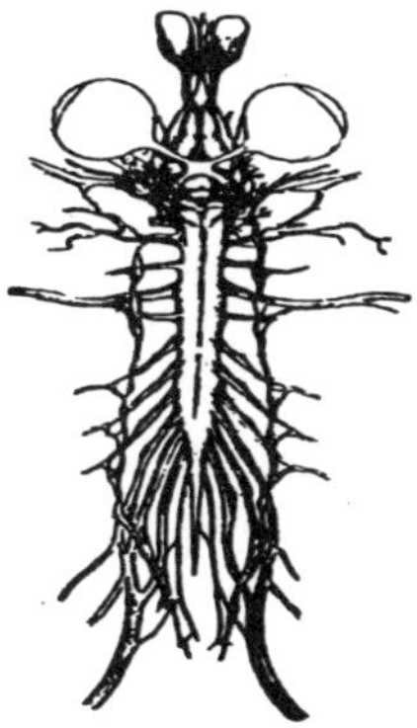

Fig. The Nervous System of A Frog in Ventral View

Other nerve-fibres carry impulses outwards and are connected with what we call effector organs—organs which are capable of active work, the muscles and the glands. The two sorts of nerves are called afferent and efferent, because of their carrying messages to and from the central nervous system respectively; or sometimes sensory and motor, because of their main functions.

The ends of the afferent fibres inside the central nervous system are branched. These branches come into contact with those of other cells, and sooner or later with the similar branches of efferent fibres. Sometimes they connect, after a few intermediate steps, with cells which send out efferent fibres in the same region of the cord. In other cases they connect with the end-branches of cells whose fibres run up and down within the central nervous system, often up to the brain; but even so the other ends of these fibres are always connected, directly or indirectly, with efferent fibres.

All messages from receptor organs, therefore, always pass to the spinal cord or brain; within these organs, they are, either directly or indirectly, passed on to efferent fibres; and so finally to muscles or glands. The result on these effector organs may be either to start or to increase their activity ('excitation'), or else to diminish or stop it ('inhibition'). In any case, by means of the nervous system, a change in the outer world or in the body itself is made to exert an effect on the working of muscles or glands, often in quite other parts of the body.

When an action is carried out thus by an effector organ as the result of a stimulus transmitted to it along nervefibres from a receptor organ, it is called a reflex action (or simply a reflex), because the stimulus travels to the central nervous system and is then, as it were, reflected outwards again along the efferent fibre; and the arrangement of organs concerned in it is called a reflex arc.

Good examples of a reflex action are the narrowing of the pupil of the eye when strong light falls upon it, and the watering of the mouth at the sight of appetizing food. In both cases the eye is the receptor organ, but in one case the effector organ is a muscle (the circular muscle of the iris or coloured portion of the eye), in the other a gland (the salivary gland).

The spinal nerves and spinal cord, taken by themselves, represent nothing but a huge system of interrelated reflex arcs; in other words, a wonderful arrangement for translating changes in the outer world, through their effects on receptor organs and nerve-fibres, into action. A particular stimulus will automatically call forth a particular action because there is a path (predetermined through heredity) in the nervous system from the receptor organ affected by the stimulus to the muscle or gland which acts. That this is so can be shown by destroying the whole brain of a frog.

The rest of the frog is now quite unconscious, but can

continue to live for many hours. The limbs hang limp; but if the toes are pinched, the leg will be drawn up; if a drop of acid be placed on the skin of the back, the leg will be raised to wipe it away. It may be noted that the number of different paths is considerably smaller than the number of afferent. The nervous system in this respect is like a funnel, with stimuli poured into its broad top, to issue in a narrower stream of action.

What then does the brain do in the frog's organism? In the first place it receives the messages from the large organs of special sense in the head—sight, hearing, smell, and taste. Secondly, it is the main controlling centre of the animal. Thirdly, it (or rather part of it) is the seat of consciousness. Owing to the first cause, the constant stimulation through the organs of special sense, the brain is in more intimate relation with the outer world, and with more of the events of the outer world, than is the spinal cord or any other part of the frog.

Then there is the question of control. In the spinal cord not all the branches of afferent fibres connect at once with branches of efferent fibres leaving the cord; some enter into relation with special fibres which run up inside the cord to the brain, and there make connexions with fibres of the brain. From other parts of the brain, fibres run down the cord again, and come into contact there with branches of efferent fibres.

The efferent nerves of the cord, therefore, can be affected first by the messages of the sense-organs brought along the afferent fibres, and secondly by messages from the brain. Now the brain, as we saw, is in better contact with the outer world than is the rest of the frog; further, it is the seat of memory. Thus the messages from the receptor organs of the skin and of the inside of the body are sent up to the brain and there brought into relation with messages from more distant surroundings and with records of the past.

In just the same way, in an army in the field, the battalions in the line send up reports to Head-quarters of what they have discovered about the enemy, and of what they themselves are doing; and at Head-quarters these reports are considered in the light of the much wider knowledge both of present and past conditions, acquired through the Intelligence and Operations branches. Finally, just as a battalion commander might think that to attack was the right policy and yet might be ordered to remain inactive owing to some situation far behind the enemy's front, of which he knew nothing, so the reflex action which would inevitably take place if the spinal cord were left to itself may be altered or entirely stopped by messages reaching it from the brain.

A sneeze, for example, is a reflex action, but we can usually stop it by an effort of will if we realise, for instance, that to make a noise will put us in danger, or remember that there is an invalid in the room who must not be awakened. Finally, states of consciousness seem to correspond with special ways of setting the connexions in the brain. When we are angry, the brain-connexions are so set that messages may run out to all the effector organs concerned in attack and defence; when we are afraid, the machinery for running away is put in readiness; when we are depressed, it means a damping-down of all our general activities, and so forth. The reflex arcs connected with the spinal cord thus represent the machinery by which an animal's actions are possible. But the particular actions carried out depend upon the way in which the brain influences this mechanism.

The range of actions possible to a frog is much less than that possible to a dog or monkey; and that of a dog or monkey, again, very much less than that possible to a man. The differences between the spinal-cord machinery of the three types of animals, however, is comparatively small; it is through differences in the brain that the same general machinery can be adapted to many more situations and made to carry out a far greater number of distinct actions.

By means of the nervous system, then, as well as through the blood-system, co-ordination is carried out. But whereas the co-ordination effected by internal secretions in the blood is primarily between one internal organ and another, that effected by nerves is largely between the outer world, as it stimulates the receptor organs, and internal organs. What is more, a much more finely adjusted co-ordination can take place through the nerves than through the blood. Actions like the accurate bringing up of the frog's leg to the particular spot on the back which is stimulated, or my using my pen to write these words, involve just the right degree of contraction of a large number of muscles, and are far more complex and nicely adjusted in detail than the secretion of the pancreas under the influence of the hormone from the intestine, or the darkening of the frog's skin when the pituitary hormone is injected.

Furthermore, in the working of the nervous system, we find that one part is in the relation of central head-quarters and so dominates or controls the rest, whereas nothing of the sort happens in co-ordination through the blood. We have not yet mentioned the reproductive system, by means of which new frogs are produced from old. So far we have dealt with the working of the frog's organism, and the plan of its structure. It remains to consider the frog in relation to its surroundings. When we do this, we find that there are so many correspondences that they cannot well be due to chance.

The common frog spends much of its time in the water; and its hind feet have a web between the toes. It uses its skin for respiration; and lives in damp places. It feeds on small animals; and has a tongue suited for seizing such prey with lightning rapidity. It is preyed upon by various birds; but has a blotched skin to camouflage it, and, furthermore, a skin which changes colour with its surroundings, becoming darker on a dark background and vice versa.

The tree-frog, on the other hand, lives among the leaves of trees; and it is coloured green. It must climb tree-trunks; and possesses special adhesive pads on its feet. The tongue of a frog would be of no service to a grass-eating cow, nor its webbed feet to a dog or cat, nor its skin to a desert-dwelling animal.

In brief, the structure and habits of the frog, like those of all other organisms, fit its surroundings; and we say that it is *adapted* to its particular environment. *How* it is that animals and plants are adapted to their surroundings is another and far more difficult problem, which we must leave for the present. Our next immediate concern is to penetrate into greater detail of the frog's anatomy. So far we have only considered structures visible with the naked eye or with the help of a hand-lens. But by means of the microscope we can submit the frog's tissues to a magnification of several hundred diameters, and see much that was invisible before. This branch of Zoology is called Histology, or the microscopic anatomy of tissues.

Perhaps the most important fact revealed by these methods is the fact that all tissues are made up of definite units of living substance, usually called cells. The name cell is not a very good one, as it suggests hollow boxes, and the living cell is not hollow and is in no sense like a box. However, the name was first given to the hollow boxes of which cork and pith are seen to be composed when examined under the microscope; these are the dead cases of once living tissueunits. Then the name was transferred to the living tissueunits of plants, living contents, box, and all; and finally, when it was clear that the slimy contents were the essential and the box something incidental, found in plants and not in animals, the term became restricted to the contents, and the box of cellulose in plants was called the cell-wall.

A cell is the functional unit of living substance or protoplasm. This semi-fluid, almost transparent material, generally containing granules, is a complex mixture of substances, some of which in their turn are also of extreme chemical complexity. Cells, both of animals and plants, normally possess a specialized cell-membrane at the surface, which regulates the passage of materials in and out of the cell, and a central usually rounded body with a definite membrane of its own, the nucleus. The protoplasm apart from the nucleus is called the cytoplasm.

All tissues are made up of living cells, together in some cases with dead substances produced by cells; and each tissue consists of one or a few characteristic kinds of cells, arranged in a characteristic way. Blood is the tissue (if we may stretch the term 'tissue' to cover a mass of cells not joined together, but moving freely in a fluid) which is most easily examined microscopically. If we look at a small drop of frog's blood under the microscope, we shall see that it contains thousands of cells. The commonest type is a flattened oval in shape, with its cytoplasm of a faint straw colour, and containing a central uncoloured nucleus. In bulk these corpuscles give blood its red colour, and are called the red corpuscles. Their colour is due to a pigment called haemoglobin which they contain, and by means of which they convey oxygen round the blood-stream. Human red corpuscles are smaller, bi-concave, and without a nucleus. Besides these, there will be seen a number of smaller, uncoloured bodies, capable of slow movements and alterations of shape; these are called the white corpuscles. Most white corpuscles have the power of devouring bacteria and other foreign particles. When they collect in large numbers inflammation is often found. They are the body's microscopic policemen and scavengers.

Next we distinguish a whole group of tissues which form linings to surfaces, whether external or internal; such a lining is called an epithelium. The body-cavity, for instance, is lined by a single layer of flattened cells fitted together like a simple jig-saw puzzle; the absorptive lining of the gut by narrow cylindrical cells; the lining of the wind-pipe by cubical cells armed with tiny lashes or cilia which beat uninterruptedly and drive any small foreign particles up and out of the wind-pipe.

The outer skin, or epidermis, on the other hand, is an epithelium of many layers of cells. The lower ones are roughly cubical, and are continually producing new cells which become gradually transformed into horny plates to be rubbed off on the outer surface as scurf; in this tissue, therefore, cells are constantly dying throughout life, being sacrificed for the good of the whole organism. The horny layer is much better developed in land-forms than in water animals or the moist-skinned frog.

Organs of touch are scattered just below the epidermis, and numerous glands open on it—sweat-glands, for instance, in man, slime-glands in the frog. *Glands* are those parts of the organism whose function it is to extract or manufacture particular substances from the blood, whether they are substances of which the organism will make use, or substances of which it must rid itself. The pancreas secretes pancreatic juice for use in digestion, the kidneys secrete urine for elimination from the system. The simplest kinds of glands are single cells, such as the mucus or slime cells which are scattered among the absorptive cells of the frog's intestine, and secrete a lubricating fluid.

Most glands, however, are manycelled tubes or pockets of epithelium, either unbranched or slightly branched like the glands of the stomach, or muchbranched like the liver or salivary gland. Their cells are usually more or less cubical. Generally little spherules of the substance which they secrete are to be seen within their cells; after a gland has been in action, however, these are seen to have disappeared—they have been discharged.

Fig. A. Section of Cartilage (From Frog's Breast-bone) Highly Magnified.

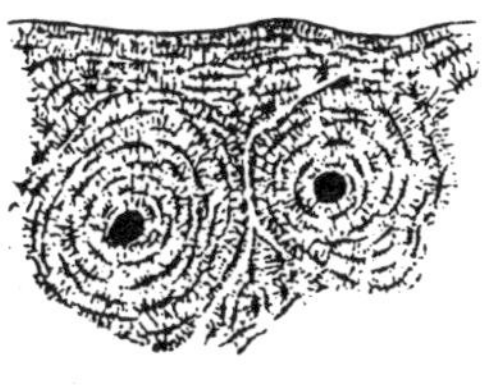

B. Section of bone (from pig's thigh-bone) highly magnified. In life, blood-vessels run in the large circular black spaces, and bone-cells occupy the smaller branched spaces. The intermediate material is the bone-matrix.

The cells of muscles are very remarkable structures. The simplest are found in smooth muscle; they are very much elongated, and are marked with a number of fine fibrils along their length. In voluntary muscles (those attached to the skeleton), each unit is made of a number of cells run together, as is evidenced by the number of nuclei which it contains. Further, in addition to longitudinal fibrils there are a number of transverse bands across the fibres of voluntary muscle. The presence of these bands is somehow associated with greater rapidity in contraction.

Next we have a group of tissues called the supporting tissues. They have the property of secreting dead substances out of their living selves, and by this means building a skeleton or framework. In gristle (cartilage), for example, a number of roundish cells can be seen, embedded in a stiff jelly-like substance which they have produced.

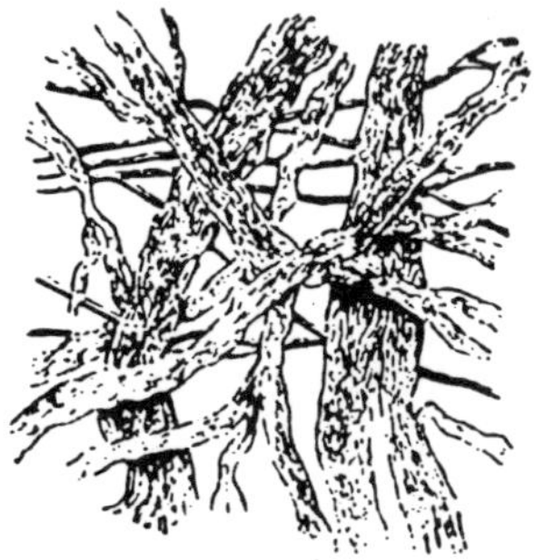

Fig. The Fibres of Ordinary Connective Tissue After Removal of the Cells

The same is true of bone, save that there the cells are usually arranged in definite systems, often concentric, and that fine branches from them penetrate the ground-substance in every direction. Connective tissue, on the other hand, which binds up every organ in a fine firm sheath of tissue, has a number of scattered cells which produce bundles of fibers, interlacing in every direction, and giving great tensile strength combined with elasticity. If we could conjure away every other tissue of the body, we should still see the outlines of every organ, including the course of every vein and artery, every nerve, preserved for us in this pervading scaffolding of connective tissue.

Nerve-cells or *neurons,* as befits their remarkable functions, undergo perhaps more change during their development than does any other type of cell. In early stages, they are irregularly rounded, like most other cells at this period. After a time, however, a prolongation grows out at one end, and one or more similar prolongations at the other. These continue to grow, and become the conducting nerve-fibres we have already spoken of; some of them may reach relatively enormous lengths, the muscles of the toes, for instance, being supplied by fibres which run all the way from cell-bodies in the spinal cord—a distance of several yards in the largest animals known.

A nerve-fibre is thus always attached to a nucleated cell-body, and cannot exist without it; if a nerve be cut, all the fibres which are no longer attached to cells die; whereas those parts of them which are still in connexion with the cell-bodies will live and regenerate. The nerve-cells in parts of the cerebral hemispheres are remarkable for the degree of branching shown by their processes; it is probable that these are concerned with memory, and in man with the association of ideas.

Fig. Diagram to Show the Main Parts of A Nerve-cell

In the same sort of way in which a substance is chemically composed of molecules, so the higher animals and plants are built of cells. They too, of course, are in their turn composed of molecules, and those of atoms; but the cells are the smallest definable *biological* units. In a sense, the body is a colony of cells. In a beehive, the lives of the individual bees are subordinated to the good of the colony; so too the individual cells are subordinate to the good of the body, but the subordination is far more thorough, and the 'colony' can act as a single whole far more efficiently than the hive.

This it can do in spite of the enormous number of cells which it contains. If human beings were blood-corpuscles, the population of London would almost fit into a cubic millimetre of human blood, and the population of the world into a dozen drops.

A frog, then, is a mass of living substance organized on a particular plan, and this plan bears a definite resemblance to that of human beings, although it is altogether different from the plan on which both crayfish and cockroach, for instance, are constructed. It is composed of cells; these joined together into tissues; these again into organs and systems of organs. Each organ is constructed so as to work in a particular way, which is normally for the good of the whole organism to which it belongs.

Some organs, like the digestive system and the glands, carry on chemical work; the skeleton is a passive support and protection; the blood-system is the go-between for all the others; the muscles by changing their shape move the whole organism or alter the state of the organs; the receptors are like windows into the world of events.

But through one window one set of events only can be seen, through another window only another set; the nervous system ensures that the reports of different kinds of events shall be co-ordinated, and that on the whole the right response shall be made to the right event.

Thus the nervous system, together with the chemical regulation effected by way of the blood-stream, makes it possible for the animal to act as a unity, as a single whole, and so to deserve the title of *organism*.

Chapter 2

Development and Heredity

Frogs grow old as well as men, and will die of old age even if they have not previously met death by some accident. If there is to be a race of frogs, there must therefore be a continual production of new frogs to take the place of the old. Our next inquiry must be into the method adopted to ensure this *reproduction*.

If we run an individual frog's history backwards through our minds, as a film can be run backwards through a cinema, we find that the full-sized or adult frog was preceded by a young frog of much smaller size but of the same general shape and structure. Before this, however, the frog was so different as to merit a different name; it was a tadpole, lived only in the water, fed mainly on vegetable food, had only rudimentary legs, and swam with a tail.

It did not breathe by lungs, but by gills; these are branched outgrowths of the body-wall with many capillaries just below a very thin skin. They grow on the outer borders of slits which lead from the cavity of the mouth to the side of the neck, and, since water is continually being sucked in at the mouth and then forced out through these gill-slits, they can be always taking up dissolved oxygen from the water and passing carbon dioxide out into it.

In large and mediumsized tadpoles, the gills and gill-slits are covered by a flap of skin or gill-cover; but in quite small ones the gills stand out free on the side of the neck. As the adult frog grew out of a smaller frog, so the large tadpole has grown from a smaller tadpole of the same general form and containing the same systems of organs, except that it is altogether without limbs. The small tadpole, in its turn, hatched out of a glutinous covering, which with several hundred others formed a mass of frog-spawn.

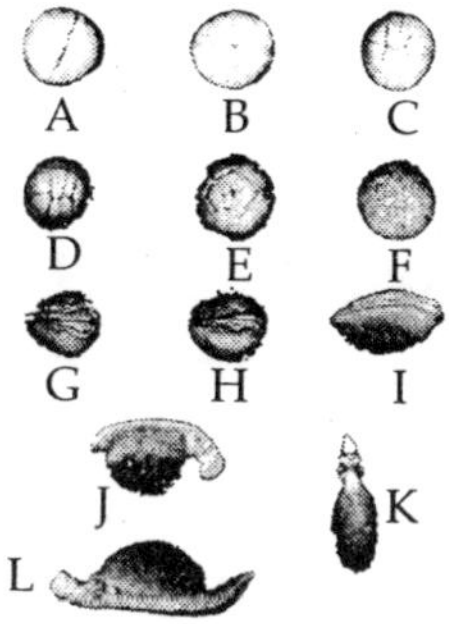

Fig. The Early Development of A Tailed Amphibian

A, egg during the division into two cells. *B*, 4-cell stage. *C*, beginning of 16-cell stage. *D*, *E*, later segmentation. *F*, blastula. *G*, neural folds have appeared. *H*, neural folds closing. *I*, neural folds closed to form neural tube. *J*, from the r. side. Head and tail sharply marked off from yolk-mass. The gill-slits are seen in the neck region, the muscle-segments on the fore-part of the trunk.

K, later stage, from the r. side and upside-down. The tail has grown, and has developed ventral and dorsal fins. Fore and hind limb-buds visible, yolk-mass relatively smaller; three tufted gills in the neck region. *L*, a similar stage, from below. In front of the gills in the middle line is the mouth, with the eye-rudiments just in front of it.

If we had gone farther back and examined the developing frog inside the jelly a few days before hatching, we should have seen that it still had the general form of a tadpole; but its gills were mere knobs, its tail not fully formed. Before that again there was a time when, although the general plan of the internal organs was the same, the organs themselves were not present as working pieces of machinery, but merely blocked out in an undifferentiated state, with their component cells not yet specialized as in the adult.

The liver at this stage, for instance, was represented merely by an unbranched tube; the heart was an S-shaped blood-vessel, as yet without muscles or valves; the chief parts of brain and spinal cord could have been seen, but in a much more rudimentary shape than in the full-grown frog, and with no nerve-fibres yet formed by outgrowth from their cells.

Still earlier, the developing animal had no particular resemblance to a tadpole, and even the main organ-systems were barely recognizable. The future brain and spinal cord, for example, were represented by a groove along the back; the digestive system was an irregular space within a mass of yolk in the interior. Before this, the organism (when passing through what is termed the gastrula stage) was quite spherical, consisting of an outer sheet of several layers of small cells, an inner sheet mainly composed of big yolk-laden cells surrounding the cavity which is the first rudiment of the gut, and a middle sheet between the other two.

A day or so earlier, in the blastula stage, there was but one sheet of cells, and no digestive rudiment. Earlier still, there were only a few large cells; and at the last, we can trace our frog back to a single very large cell, black above and white below, loaded with reserve food in the shape of yolk, and containing but a single nucleus. This is the fertilized egg. 'What is the origin of the fertilized egg?' is the next obvious question. But before attempting to answer this, we must go in somewhat more detail into the actual progress of development.

Fig. Transverse Microscopical Section Across Very Young Cat Embryo in the Germ-layer Stage.

This can be divided into a number of periods. The egg contains plenty of yolk, in order to provide food for the young frog before it can feed for itself. It is consequently a very large cell, and the first step to be taken is to divide it into cells of a more convenient size—the bricks to be used in the future building. The egg divides into two, these into four, and so on, until the blastula stage is reached. This marks the close of the first period, usually called the period of segmentation. The second period ends in the formation of a rough—a very rough—ground-plan of the future organism. A fold of the dark, smaller cells grows over the larger yolky cells, and the crack between the inner layer of the fold and the yolk afterwards swells out to be the first rudiment of the digestive system.

If you hold one end of a sheet of paper in place and bring the other end down towards the first, you will produce a similar fold; but the movement of the living fold is caused by rapid growth of cells just about the folded part. Between the inner and outer layers of the fold a third layer of cells is split off. Now the first groundplan is ready—there are three distinct layers of cells in exis- tence.

The outer layer will later give rise to the epidermis of the skin, the sense-organs, and the nervous system; the inner layer to the gut and all its appendages such as liver, pancreas, thyroid, lungs, and gill-slits; and the middle layer to the muscles, the skeleton, the connective tissue, the bloodsystem, the kidneys, and the reproductive organs. These three primary layers are called germ-layers, so that this is the period of germ-layer formation. Segmentation and germlayer formation are a good deal simpler in forms with less yolk in their eggs.

In the third period a great advance is made—the main systems of organs are blocked out. The embryo lengthens: a groove forms on the back, deepens, and its sides arch over and meet to produce a tube; this tube is the rudiment of the whole central nervous system, and of most of the nerves.

Tiny pits appear on the side of the head, representing the future nose and ear; and from near the front of the nervetube, two hollow outgrowths arise which will form the main part of the eyes. Below the nerve-tube, a long straight rod is nipped off from the top of the gut. This is the notochord, the early and less complicated precursor of the true backbone.

On each side of the notochord, the middle layer of cells grows rapidly, and cuts itself up into a series of blocks of tissue, the muscle-segments; from these the voluntary muscles will be formed. Below them, a split appears in the middle layer—

the rudiment of the body-cavity. From its walls a series of funnel-shaped tubes grow out, and their ends unite and form a duct which grows back to open at the hind-end—the rudiments of kidneys and of ureters.

The mouth and anus are pierced; outgrowths of the gut produce the pancreas- and liver-rudiments; other outgrowths reach the exterior and will form the gill-slits. Scattered cells of the middle layer unite to form tubes, and these tubes join up to give the blood-vessels. The main characteristic of the period has been the formation, in a very short time (not more than 48 hours at ordinary temperatures), of a great deal of visible structure. We may call this the period of primary differentiation of organs.

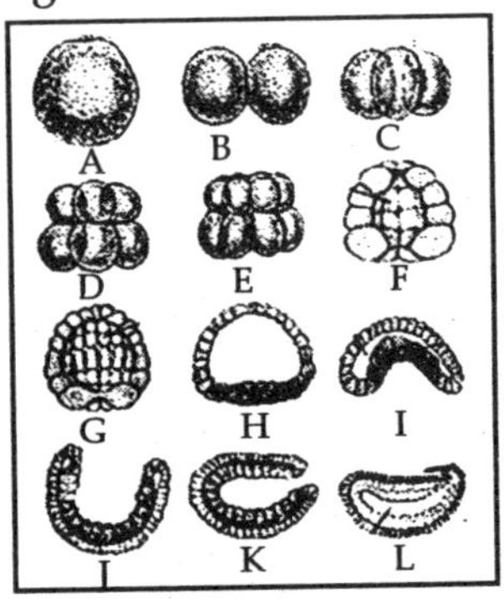

Fig. The Early Development of Amphioxus.

A, fertilized egg, unsegmented. *B,* 2-cell stage. *C,* 4-cell stage. *D,* 8-cell stage. *E,* 16-cell stage. *F,* early blastula (32 cells). *G,* late blastula. *H,* flattening of vegetative side of blastula. *I,* invagination to gastrula. *J,* partial gastrulation. *K,* late gastrula. *L,* beginning of formation of neural tube. *ach,* primitive gut. *bc,* blastocoel. *bp,* blastopore. *ec,* ectoderm. *en,* endoderm. *mac,* larger vegetative cells. *mic,* smaller animal cells. *np,* neural plate beginning to form neural tube. *pb,* polar bodies. At the end of this period, the laying out of the detailed ground-plan, some little time before hatching, the chief organ-systems and organs are thus all present, but they are not yet in working order—their cells are still all of the same general type, not specialized for their various duties. The next, or fourth period, therefore, is primarily chiefly that of the differentiation of the tissues. The gut-rudiment, growing capable of digesting, becomes a gut; the muscle-rudiments become muscles; the nerve-tube becomes a brain, spinal cord, and nerves; and so on with the rest of the organs of the body.

This change is accomplished by the time of hatching for some organs, a little later for others. It takes place through transformation of the shape and structure of cells. In the next period the animal *works*, and becomes selfsupporting; feeding and consequent growth are its chief characteristics; but towards its close, preparation is made for the adult stage by the appearance of limbs and of lungs-again first as mere rudiments, later as organs capable of working.

This is the *larval* period, a larva being a stage in an animal's history when it is self-supporting but (and therefore no longer an embryo) radically different in structure and mode of life from the adult. The sixth stage, or period of metamorphosis, is a violent transition from larva to adult, from water to land. If we could not actually follow the transformation, it would be impossible to guess that a young frog and a tadpole were stages in one and the same animal's life.

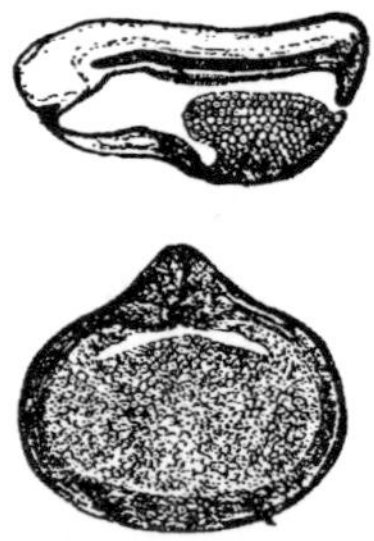

Fig. A Longitudinal Vertical and a Transverse Section Through a Late Embryo of a Frog

The limbs grow rapidly, the shape of head and trunk and the colour of the skin become altered, and there are internal alterations, such as a shortening of the intestine in view of the change from vegetable to animal diet, the remodelling of the skull, and the replacement of much cartilage by bone.

But the most remarkable changes are those affecting the

tail and gills. These do not drop off, as is still believed by some unobservant people; they are absorbed. They have been built up; now they are unbuilt. Their tissues lose their differentiation, degenerate, and are used as food-material. Thus the development of an organ need not always be forwards; it may be reversed. After the change, another growth-stage sets in, which we may call the juvenile period, when the animal is definitely a frog, but not yet grown up. After a time, however, growth slows down, and at about the same time sexual maturity begins. So the eighth of our periods, the adult period, starts. This is the longest of all the periods, and is best developed in the highest forms of life.

It is in them a condition in which neither growth nor breakdown has the upper hand, a period of balanced activity, which has been compared to the apparent rest of a 'sleeping' top. Frogs, like mammals and birds, have a well-marked adult period in which growth is absent. Many lower animals, however, such as most crabs and shell-fish, go on growing throughout life, and in them sexual maturity, not cessation of growth, is the only criterion of the adult phase.

Finally, however, the ninth period sets in old age. The metabolism grows feebler, the organs no longer work so well, and even if a violent end does not terminate the animal's existence, as is almost always the case in nature, its life eventually comes to death as an inevitable close.

Such a death is unavoidable and normal; we may call it natural death a real wearing-out of the tissues, and a consequent collapse of the organism like the collapse of the hundredyear-old 'one-hoss shay' in the poem.

The development of an animal like the frog, then, consists partly of growth; partly of differentiation or increase of complexity; thirdly, of metamorphosis from one form to another; fourthly, of the attainment of a relatively stable, balanced state, in which growth and differentiation are almost absent; and lastly, of the loss of this stability, the wearing-out of the machine old age, and death. At *A*, a medium-sized oocyte (miniature ovum) surrounded by a follicle one cell layer thick.

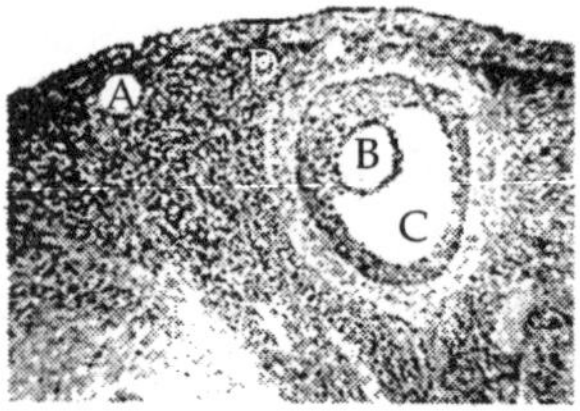

Fig. Micro-photograph (× 150) of a Section Through the Ovary of a Mammal (Cat)

At *B*, a larger oocyte; its follicle has become several layers thick, and a cavity containing fluid has been formed in it. In the oocytes at *A* and *B* the large nucleus can be seen. At *C*, a very young oocyte. *D*, the edge of the ovary, bounded by a layer of germinal epithelium. Now having traced the development of a grown-up frog from a single fertilized egg, we can return to our question as to the origin of this fertilized egg. If we look at the ovaries or reproductive organs of a female frog in early spring, we shall find in them, and actually produced by them, a great number—one to two thousand—of eggs *apparently* similar to fertilized eggs; a little later in the season, these will be found loose in the swollen lower part of the oviducts, each surrounded with a thin layer of jelly. But if we take these eggs and put them in water, they will not develop.

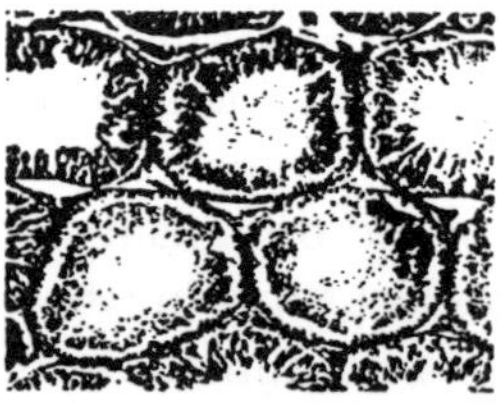

Fig. Section Across the Testis of a Mammal (Rat)

For them to start development, fertilization is normally necessary; and by fertilization we mean union of the egg with a reproductive cell of the male.

In all animals, the eggs are cells detached from these special organs of the body, the ovaries. The reproductive organs of the male are the testes, small whitish bodies consisting mainly of a number of microscopic tubes.

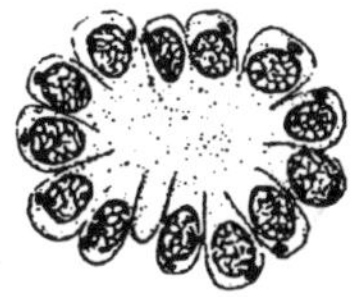

Above, a group of spermatids, produced at the close of the maturation divisions. Each has a nucleus and a centrosome. (In the worm, the cell-bodies are not at this stage fully cut off from a central mass of protoplasm.)

Centre, a similar group showing the first stages in the elongation of the spermatids, their outgrowth at one end, and the elongation and condensation of their nuclei.

Below, left, three nearly mature sperms. The outgrowth has become converted into the vibratile 'tail' (flagellum). There is still a distinct layer of cytoplasm over the nucleus.

Below, right, a mature sperm. The tail is further elongated, and takes origin in a 'middle-piece', containing the centrosome. Then comes the 'head', consisting of the much-elongated nucleus, surrounded with a mere film of cytoplasm, and tipped with an organ which facilitates the sperm's entry into the egg.

Fig. Four Stages in the Differentiation of Spermatozoa, From the Earth-worm.

The cells forming the walls of these tubes divide, and some of the fresh cells thus produced change their shape in a remarkable way; the nucleus becomes long and dense, and the rest of the cell becomes transformed into a sharp point at one end and a swimming tail at the other.

They are now ripe, and are called spermatozoa or sperms. These are shed into the hollow of the tubes, pass down canals into and through the kidney, and down the ureter, to be stored in a pouch on its side. A drop of the contents of this taken in early April and mixed with water will show thousands of sperms in active movement; a male frog produces billions in a season. When the female lays her eggs, the male sheds his sperm over them. The sperms are attracted by the eggs, and start burrowing into them. Once one has succeeded in forcing its nucleus inside, a change takes place over the egg's surface which prevents any further spermatozoa from entering. The sperm-nucleus then swells up, and becomes very like the nucleus of the egg.

The two nuclei meet, and actually unite to form one. This completes the act of fertilization; and only after this does development begin. Sexual reproduction in the frog then consists in this: that union of two cells, or at least of the nuclei of two cells, takes place, that the cells at the moment of union are detached and independent, but previously formed part of the bodies of other individuals of the species; and that the cell formed by their union develops into a new individual.

It has been found possible with the eggs of frogs, worms, sea-urchins, and other animals, to make the egg start its development by artificial means (by chemical treatment in sea-urchins; by heat in starfish; by pricking the egg with a fine needle dipped in blood, in frogs), without any sperm being present, so producing fatherless animals. Some of these have been raised to maturity, and appear perfectly normal. Thus fertilization consists really of two separate and separable processes—activation, or the starting-off of the egg on development, and the union of the nuclei of male and female reproductive cells.

The two cells which thus unite are called *gametes* or 'marrying cells'; and the product of their union is called a *zygote*—something formed by the 'yoking together' of two gametes. Usually the gametes are, as in the frog, of two markedly different kinds, the female gamete large and yolk-laden, the male gamete tiny and active.

Fig. Human Unfertilized Egg (Oocyte) and Sperms, on the Same Scale

The egg is seen surrounded by its membrane and this again by protective and nutritive follicle cells (*FC*). It contains a large nucleus (*N*), which has not yet undergone reduction, and some yolk grains (*Y*).

The nucleus contains a nucleolus. Below, at *B*, are two sperms, the lower one in profile. The 'head' contains all the nuclear material, condensed. The ovum is about 2001 (0.2 mm.) in diameter. This difference between the two gametes, however, is not universal. In many low forms of life, both plant and animal, the two are alike; and in some, as in the single-celled animal Paramecium, only nuclei and not whole cells fuse with each other.

(*a*) Immature (unfertilized) egg with nucleus in 'resting' phase.

(*b*) Fertilization has just occurred.
The nuclei of egg and sperm are approaching each other.
The chromosomes have begun to appear.

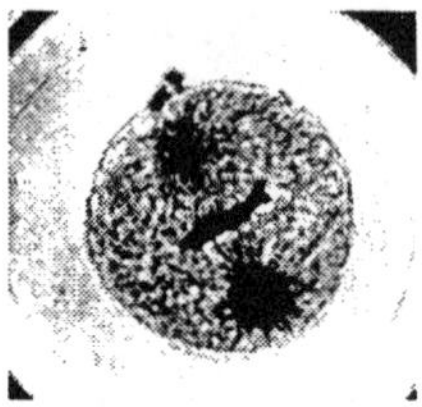

(*c*) Side view of the equatorial plate stage of the first division of the fertilized egg. The spindle is clearly seen with the centrosomes and asters at its two ends. The chromosomes have arranged themselves round its equator.

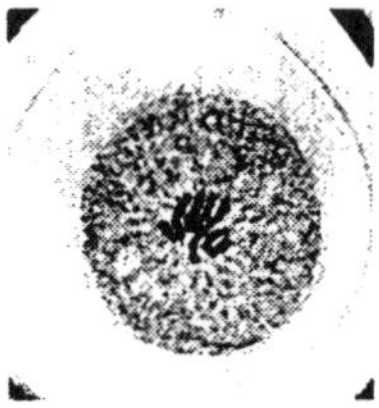

(*d*) End view of the same stage.
The four chromosomes are clearly visible.

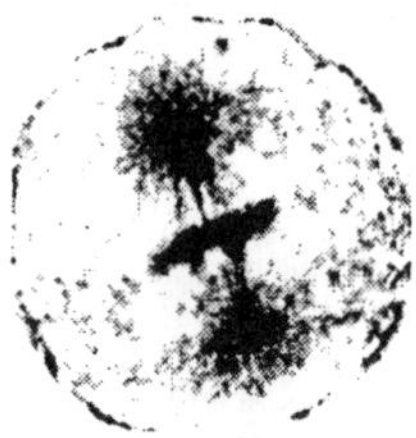

(*e*) Side view of a slightly later stage.
The chromosomes have now split longitudinally.

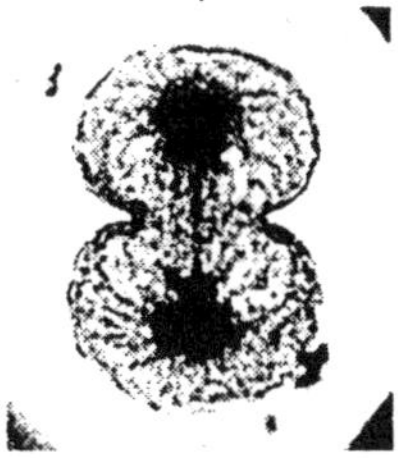

(*g*) 'Telophase' stage. The two sets of chromosomes have moved apart to the two asters; the cell is deeply constricted.

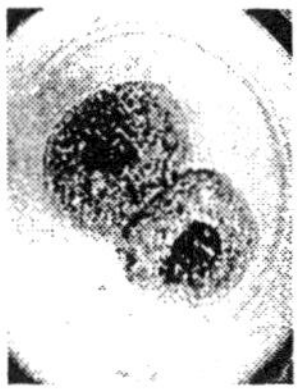

(*h*) Two-cell stage. The egg has completely divided into two cells; in either cell the chromosomes have joined up to form a 'resting' nucleus.

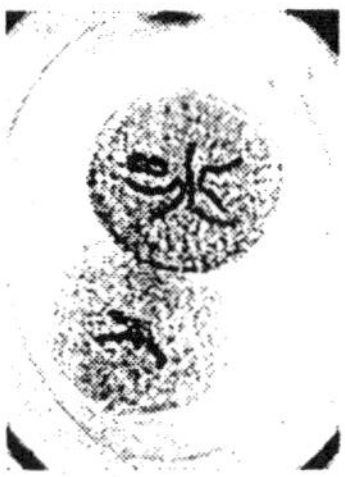

(*i*) Beginning of second cleavage.

Fig. Mitosis, the fertilized egg of the round-worm *Ascaris megalocephala,* from the intestine of the horse, which possesses two pairs of chromosomes.

Thus the essential fact in sexual fusion is the union of the two nuclei. The difference between the two gametes is only secondary; the size and yolk-contents of the egg serve to give the developing embryo a good start in life after fertilization; and the shape and activity of the sperm ensure that the two gametes shall meet.

The race of frogs, then, can be thought of as consisting of

a number of continuous streams of living substance. The streams flow for the most part in the form of zygotes, starting as fertilized eggs and developing into mature frogs; but for some of the time they subdivide to flow in the form of gametes, and then such of these streams as do not die out, unite in pairs to become zygotes once more or, to put it in another way, the species *Rana temporaria,* or any other kind of higher animal, such as the human species, comprises not *two* but *four* kinds of individuals—male and female zygotes, and male and female gametes. The zygotes are large and long-lived, the gametes small, short-lived, but far more numerous.

To understand fully what this implies, we must again turn back. We have said that cells divide, but have not described the process. As a matter of fact, it is a very remarkable one. When a cell is ready to divide, the wall of the nucleus breaks down, and its contents change from their chiefly fluid state into a number of more solid rod- or strap-shaped bodies, which, because they become coloured by many dyes, are called *chromosomes.*

Meanwhile, a starshaped figure of radiating fibres is seen in the cell. This divides into two, forming a spindle-shaped set of fibres with a radiating 'star' at each end, and the chromosomes arrange themselves where the fibres from the two stars meet, in the centre of the spindle. The chromosomes next split right down their length, and the halves travel away from each other, one to one pole, the other to the other. Finally, the body of the cell divides, the set of chromosomes at either pole swells up to form a nucleus again, and there are two cells, each with its nucleus, in place of one.

This process of nuclear division (called *mitosis,* on account of the thread-like chromosomes, from the thread) ensures that each chromosome is divided exactly along its length. The units which determine the hereditary characters of an animal probably lie along the chromosomes, so that as a result of mitosis a complete set of halved units is distributed to both of the two cells arising at any division; and these halved units, since they are alive, soon all grow up to their original size again. The number of chromosomes is always the same for a given race of animals or plants; not only this, but often the individual chromosomes can be distinguished from each other by differences of size and shape; and when this is so, they can be arranged in a series of pairs, so that the total number is always even. In man, for instance, the total is 48; in the fruit-fly, *Drosophila melanogaster*, it is 8; in the Mexican salamander 28; in Ascaris 4; and so on.

At one of the cell-divisions just before the formation of gametes, however, the process is quite different. Instead of the chromosomes dividing, the members of a pair come to lie side by side; and at division one whole chromosome of a pair is separated from the other. This process is called the *reduction* of the chromosomes, for owing to it, each of the two cells produced at this division possess only half the ordinary number of chromosomes for the species.

Accordingly, gametes have half the number of chromosomes found in zygotes; and the reason the chromosomes of the zygotes exist in pairs is that one member of the pair has come from the father and one from the mother. Thus, the ordinary higher animal or plant has two complete sets of chromosomes, like two packs of cards. This is of great importance in the study of heredity, or, in other words, of the way in which characteristics are handed on from one generation to the next.

In the centre, general scheme showing two maturation divisions, producing four gametes. The first maturation division is also the reduction division, the $2n$ or diploid number of chromosomes found in the bodycells being reduced to n, the haploid number, through the separation of corresponding whole chromosomes from each other. On the left, more detailed diagram of the process in the male, leading to the formation of four functional sperms.

On the right, more detailed diagram of the process in the female. Here, in order to retain the large size of the egg, three of every four gametes produced are minute and non-functional, and are called polar bodies, while only one becomes a functional ovum. The chromosomes derived from the animal's father are represented in black, those from its mother in white. Note that there are two of each kind of chromosome. A frog gives rise to new frogs, not to toads or lizards; within a single species, such as man, many characters 'run in families'—physical characters like black hair, or mental characters like musical talent; and it is a common observation that children inherit traits from their parents and from remoter ancestors.

The facts of heredity are often obvious enough. But what is the machinery by which they are brought about? How is it that the tiny, simple-looking egg has the power of developing into the complicated adult, and an adult just of one particular kind?

How is it possible that of two eggs, which would be practically indistinguishable under the microscope, one is predestined to produce a redhaired tall boy, the other a dark-haired medium-sized girl?

There is a great deal of evidence to support the idea that inheritance—the transmission of qualities or characters from one generation to the next—depends upon the actual transmission of small units of living substance in the gametes. These units are called the *factors of heredity*, or sometimes still more shortly the *genes*; and there is further evidence that these factors are contained in the chromosomes.

Certain results flow from this—results, for instance, concerning the numerical proportions of different types to be expected on crossing different strains; and, indeed, all the subject-matter of the comparatively new science called Mendelism, after the Austrian Abbot Mendel, who first discovered and interpreted the essential facts. Here we are concerned only with some of the more general aspects of the question.

A sexually reproducing animal or plant grows up from a fertilized egg into its adult condition. The fact that it grows up in its own particular way, that a newt's egg will produce a newt and a frog's egg a frog, although they are scarcely distinguishable and although they grow up in the same pond, must depend on the eggs containing something —let us call it the hereditary constitution—which determines the way it shall grow up.

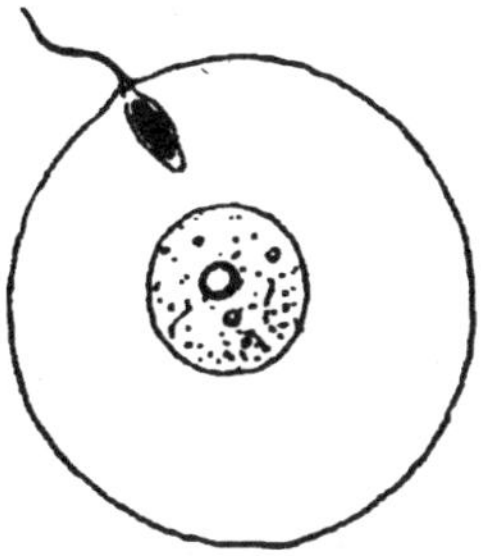

A, the sperm has penetrated the ovum.

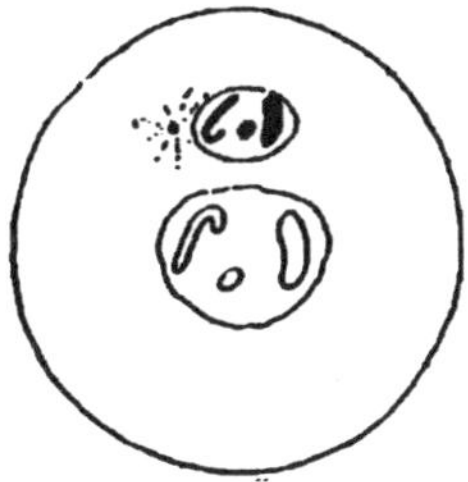

B, the sperm nucleus is swelling up, the chromosomes are appearing in both nuclei (purely diagrammatic) of the haploid number ($n = 3$) in either. The beginning of the spindle is being produced by the sperm.

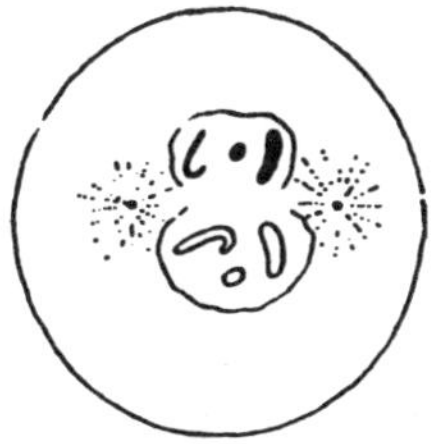

C, fusion of nuclei of sperm and ovum.

D, the spindle is fully formed, and all six ($2n$) chromosomes are arranged on its equator.

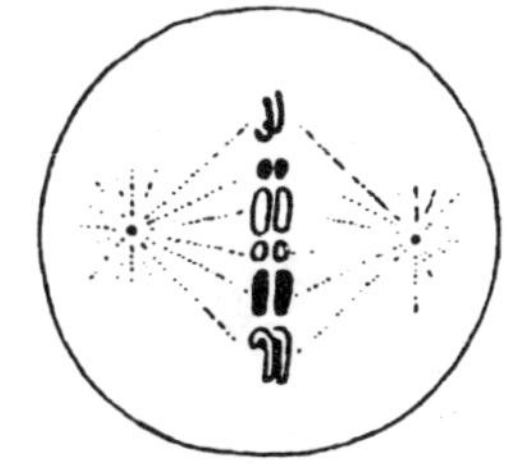

E, all the chromosomes have split longitudinally.

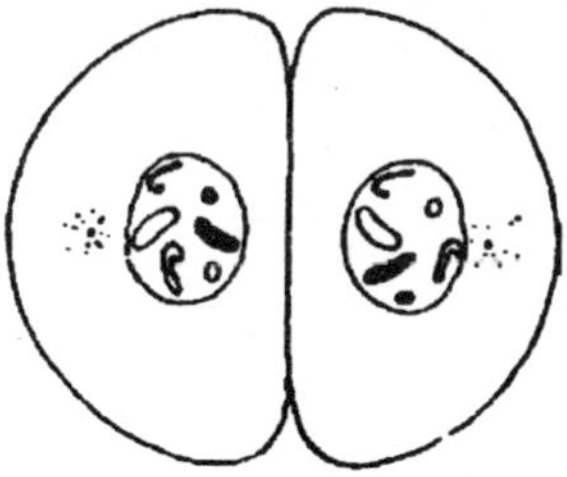

F, the fertilized egg (zygote) has divided into two cells, each with the diploid ($2n$) number of chromosomes, one set of three (n) derived from the father in the sperm, the other three from the mother in the egg.

Fig. Diagram of Fertilization

Mendel discovered two important laws concerning the way in which the hereditary constitution of an animal (or plant) is related to that of its parents. The first is usually called the *law of segregation*. This implies that the hereditary constitution is composed of a number of self-reproducing units, presumably of a chemical nature, the *factors* or *genes* mentioned above.

Normally there are two of each kind of unit in the body; but before gamete-formation, the two members of such a pair segregate, or separate from each other, so that each gamete has one member only of each kind of unit. The second law is that of *independent recombination* of units.

It sums up the fact, elicited by breeding-tests, that the segregation of members of different kinds of units is usually quite independent. It is easy to see that this is exactly what would be expected if the units were contained within the chromosomes; and, as a matter of fact, this has now been proved to be the case. The adult organism, whether human being, fly, rabbit, or pea-plant, has two complete sets of chromosomes. The two members of a pair separate from each other at reduction, so that each gamete receives one complete set; recent observations have further shown that the way in which one chromosome-pair separates at reduction is quite independent of other pairs. Thus, the single set of chromosomes in one gamete need not all have come in from the father, or all from the mother; there must merely be one of each kind of chromosome present.

If we compare the chromosomes to cards, fertilization is like the mixing-up of two packs of cards with different coloured backs. Reduction ensures that this double pack shall be sorted out into two single packs again, but pays no attention to the colours of the cards' backs. Each sorted-out pack will be complete in having one card of each kind, but any of the cards may have either a red or a blue back. Thus it will be most unlikely that, as regards the precise combination of red and blue backs, any of the sorted-out packs will be identical with any other, since the number of possible combinations is so enormous.

Each chromosome appears to contain a large number of units, and the segregation of two units will only be quite independent when they lie in different chromosomes. When they are both in the same chromosome they tend to stick together more often than they separate.

The units are the really important things in the hereditary constitution, and the chromosomes appear to be merely convenient lengths, so to speak, into which the hereditary constitution is cut up. Accordingly, the cards in our simile should really be taken to represent the separate units, and not the compound units we call chromosomes, which would really be more like suits.

If we do this, however, we must add one more point to make our analogy complete. We must suppose that any particular card need not always have precisely the same value, but that you could have one ten of hearts, say, that was a little above par, and another a little below par. In terms of factors, this would mean that although a given kind of factor always exerted the same general kind of effect in development, yet it might exist under different forms, with slight differences in effect. For instance, we know that one factor concerned with colour-production in rabbits may exist in at least four forms, one producing no pigment at all, as in albinos; the next allowing colour to appear in the ears and muzzle and other extremities, as in the Himalayan breed; the next giving the Chinchilla coat; and the last allowing full development of pigment. The different forms of one factor are called *allelomorphs*.

So far we have dealt with the generalities of Mendelian theory. They will become much clearer if we look at a couple of examples. If a certain breed of black fowl is crossed with another strain which is white with a few black markings, the offspring will all be unlike either parent, of a type known as the Blue Andalusian, which is a bluish or diluted black colour, with white 'lacing' on the feathers.

Poultry-fanciers have long tried to get a pure breed of these Andalusians by mating them with each other, but always without success—the birds 'threw' blacks and whites as well as producing more blues. When the subject came to be accurately worked out it was found that when Andalusians are mated with each other, on an average 25 per cent. of their offspring are blacks, 25 per cent. whites, and 50 per cent. Andalusian again. Not only this, but the blacks bred true when crossed with each other, and so did the whites, but the Andalusians always gave the same proportions of blacks, whites, and Andalusians once more.

Fig. A. The Results of Crossing two Pure-bred Strains of Fowls, Splashed-white and Black

P_1, the parents.

F_1, the first hybrid generation, all individuals of which are alike, of a bluish-black shade.

F_2, the second hybrid generation, derived by mating F_1 individuals together. Segregation is here shown, there being on the average one-quarter splashed-white like the splashed-white parent, one-quarter black like the black parent, and one-half blue like the F_1. These results are at once cleared up if we suppose that the black fowls carry factors which make their chickens grow up black, the white fowls factors which make theirs grow up white.

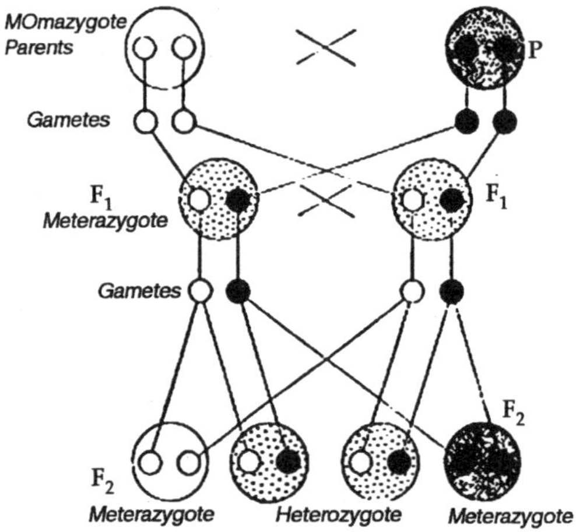

Fig. Diagram to Show How The Mendelian Conception of Hereditary Units

The organisms are represented as large circles, the genes as small circles within them. In the gametes, only the genes are represented. Unshaded represents splashed-white; black represents the gene for black, and close dotting visible black; sparse dotting represents visible blue.

When a black cock is crossed with a white hen all his gametes (the sperms) contain one factor for black, all hers (the unfertilized eggs) one factor for white. The offspring will have one of either factor; and their interaction will bring about the so-called blue colour. Exactly the same result will happen if a white cock is crossed with a black hen. When a blue cock and hen are bred together the gametes, whether male or female, will contain *either* a factor for black, *or* a factor for white, but not both. Let us call the factor for black *B*, and that for white *b*. A *B*-bearing female gamete may be fertilized equally well by a *B*-bearing or a *b*-bearing gamete; and similarly for a *b*-bearing female gamete. Thus, four possible combinations of gametes (as regards *B* and *b*) may occur at fertilization, and, on the theory of chances, will occur on the average equally of ten:

(1) *B*-bearing by *B*-bearing. (3) *b*-bearing by *B*-bearing.
(2) *B*-bearing by *b*-bearing. (4) *b*-bearing by *b*-bearing.

(1) Will give a chick with two black factors, therefore black; (2) and (3) will give Andalusian chicks once more; 3. Will give white chicks. Not only this, but out of every hundred chicks on the average one-quarter will be black, one-half Andalusian, and the last quarter white. This is made clearer by the diagram.

According to Mendel's second law, factors for different characters are inherited independently of each other. For instance, to take fowls once more, the ordinary type of comb, called single, depends on one form of a particular factor; that called rose, which is low and long, upon another.

When a cross is made between birds with these two sorts of comb, the hybrid is indistinguishable in appearance from the pure 'rose' stock. This introduces us to the fact that one form of a factor may mask another, and be what is called *dominant* to it.

But the factor for single-comb is still present in the hybrids, as is shown by breeding-tests. For whereas the original pure-bred rose-comb stock gives nothing but rose-combs, the hybrids crossed with each other give 25 per cent. pure-breeding singles, 25 per cent pure-breeding rose-combs, and 50 per cent. hybrid rose-comb birds that again do not breed true. That is to say, the difference between single- and rose-comb depends upon a difference between two forms of the same hereditary factor.

Now, if we cross a black bird with a single-comb and a

white bird with a rose-comb all the offspring will clearly be blue Andalusian in colour, with rose-combs.

But we might expect either of two things to happen in the second generation. Either the factors for black and single will stick together when the first generation forms its gametes; and then the only pure-breeding types we shall get will be black single-combs and white rose-combs again, like the grandparents.

Or else the factors can separate; if so, then we shall get gametes containing the factor for white with that for single-comb, and those containing factors for black and for rose-comb, as well as the other two types.

In this case, in addition to types which will not breed true, we should get four pure-breeding types in the second generation—not only blacks with single-combs and whites with rose-combs like the grandparents, but also two new types, blacks with rose-combs and whites with single-combs.

And this is what actually happens. This again is most simply shown in a diagram. Let us call s the factor for single-comb, S the other form of the factor, responsible for rose-comb. Then the results are as follows:

parents		*BB. ss*	x	*bb. SS*		
		(black, single)		(white, rose)		
parents' gametes		*B. s*	x	*b. S*		
1st generation			*Bb. Ss*	(Andalusian rose, but not		
				breeding true).		
1st generations' gametes						
		B. S		*B. S*		in equal
		B. s		*B. s*		pro-
		b. S		*b. S*		portions.
		b. s		*b. s*		

Each type of gamete has an equal chance of fertilizing any

other type of gamete, so that we shall get in the second generation the following pure-breeding types:

BB.SS (black, rose-comb);
BB.ss (black, single-comb);
bb.SS (white, rose-comb);
bb.ss (white, single-comb);

as well as many other types such as *Bb.Ss, Bb.SS, or BB.Ss,* which will not breed true.

The fact of the independent segregation of different pairs of factors is of the greatest practical as well as theoretical importance, for it enables us to take to pieces the hereditary constitution, so to speak, and put it together again in new ways.

More definitely, it enables us to take one particular Mendelian character which happens to be desirable in an otherwise undesirable strain or breed, and to combine it with other desirable characters in another breed.

A great deal of Mendelian work has been done with the tiny fruit-fly Drosophila, of which over ten million have been bred by Morgan in America in pedigreed cultures. Not all factor-differences produce such big or such clearcut differences in appearance as we have been considering.

Sometimes the effects are very small, or the effect of one factor interacts with that of another, and the analysis becomes correspondingly more difficult. But the principles discovered by Mendel remain the same—the hereditary constitution consists of separate units, and these segregate at reduction. This discovery is on the biological plane somewhat comparable to the discovery of the atomic structure of matter.

One point concerning human heredity deserves mention. People are often puzzled at the great diversity of physique and temperament and ability between members of the same family, although born of the same parents and brought up in the same home.

This, however, was only difficult to understand before we had Mendel's principles to guide us. For such diversity is exactly what is to be prophesied if there is independent segregation of self-perpetuating factors.

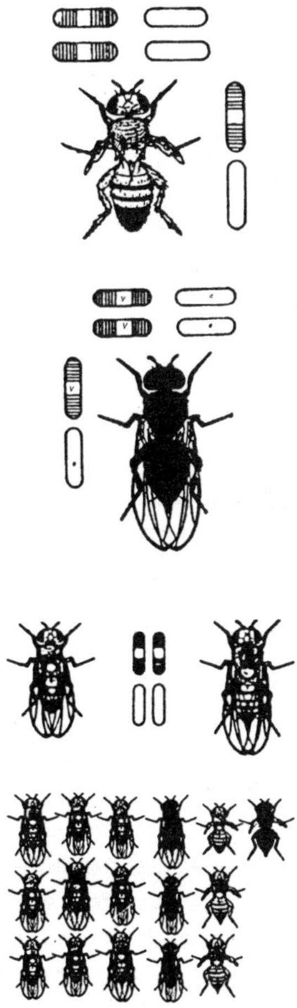

Fig. Mendel's Second Law, by Means of a Cross Between two Strains of the Fruit-fly

It is extremely unlikely that two human gametes, even from the same parents, will receive an identical outfit of factors at reduction; and even if the mother were to achieve the improbable result of producing two identically constituted ova, and the father two sperms whose constitutions were alike, there would again be enormous odds (many million to one) against the two similar sperms, among their huge crowd of competitors, happening to fertilize the two similar eggs.

On the other hand, if a fertilized ovum were to divide into two, and each part developed into a whole organism, we should expect both to be very similar, since both would possess the same hereditary chromosome outfit. This apparently is what happens in the production of so-called 'identical' twins, which as we know are often so alike as to be indistinguishable.

Another difficulty often raised is the fact that genius usually is not inherited. Again, however, this is to be expected on Mendelian principles.

If, as is probable, genius is the result of a very unusual hereditary combination of factors, then at the formation of gametes this combination will be inevitably taken to pieces again. It should not be forgotten, however, that when we take the averages of large numbers these irregularities become smoothed out, and we find a strong average resemblance, due to heredity, between parent and offspring, or between brothers and sisters.

The science of biometry deals with studies of this sort, and is very important when we are dealing with heredity in human beings, as in them, of course, experimental breeding is impossible, and the collection and handling of statistics is often the best method available.

It appears, then, that the hereditary factors not only are contained in the chromosomes, but that each one is always situated in a particular chromosome, and in a particular spot in that chromosome. Thus, each chromosome is different from every other, and each contains a great number of these factors arranged along it in a definite order, each factor always keeping between the same two neighbour-factors. Further, it appears that by far the greater part of the factors that determine inheritance from one generation to the next are situated in this way in the chromosomes.

Thus, the chromosomes are, or at least contain, the hereditary constitution of the animal or plant, and all the chromosomes taken together, from this point of view, constitute one great unit, composed of a very large number of smaller units— the factors—which are arranged within the large unit in definite positions, and in definite proportions.

The whole is like a gigantic single organic chemical compound, since the molecules of such a chemical compound are all made up of smaller parts—the side-chains and radicals and single atoms. Here, however, each of the constituent units is alive, and must have the power of self-reproduction; thus, the linking of the units together in a definite way ensures the constancy of the whole.

Fig. A Pair of Identical Twins

If there had been an excess multiplication of one of the factors during the resting period of a cell, even so, when the cell came to divide and the chromosomes to form, only the same relative amount of the factor could become linked up with the other factors of its chromosome. So the chromosomes are not only a selfmultiplying but also a self-regulating piece of machinery.

In ordinary inheritance, then, we have the handing down from parent to offspring of chromosome outfits, or, in other words, of sets of hereditary factors; and it is on these that the constancy of a species depends. They decide that a fowl's egg shall become a fowl, and a duck's egg a duck, even though they are hatched in the same incubator. They, reacting with the environment, determine what the adult organism shall be.

Apparently, however, slight changes sometimes occur in the factors, usually in a single one at one time: these changes are called *mutations*. Whereas an animal may normally have red eyes or a grey body-colour, mutations may arise which cause it, even though brought up in the same conditions, to have pink eyes or a black body-colour. (These particular mutations have actually been found in the little fly Drosophila.) When mutations occur, they are inherited in Mendelian fashion; the separation of pairs of chromosomes at reduction, and the subsequent meeting of various sets at random in fertilization, shuffles and recombines in every possible way any mutations that do occur. Thus, while the chromosomes in general act as a regulator of the constancy of the race, the occurrence of mutations and their subsequent combination in new ways by sexual reproduction provide the possibility of change. The idea of separate factors is important, for it means that when we are dealing with an animal's hereditary constitution we have not got to take it as an undivided whole, but as something which can be analysed into separate units, just as a chemical substance can be analysed into different elements in definite proportions and arrangement.

With this in mind we can turn to another question. We have discovered in outline how individuals are connected with and related to other individuals of the same species; now we must ask whether whole species are not perhaps connected with or related to other whole species.

As Charles Darwin first clearly pointed out, there is always a struggle for existence going on in nature. From each pair of frogs that breeds, over a thousand fertilized eggs are produced every season; and yet the number of frogs does not increase from year to year. There is, in fact, in every organism an over-production of young, and consequently competition. Since this takes place in all species, there is what we may metaphorically call a pressure of life; the demands upon space and subsistence are greater than the supply, and in every generation only a few of the young produced can reach maturity.

There is further the fact of the hereditary transmission of characters, and the origin of new characters by mutation and recombination. In other words, organisms are continually varying, and some at least of the new variations can be inherited. As an inevitable consequence, if a new inheritable variation appears which is of any advantage whatever in the struggle it will help its possessor to survive, and so will on the average be transmitted to later generations more often than other variations which are useless or in any way harmful. The new variation may be directly useful in competition with the other members of the species, as for instance if it made its possessor more easily able to secure its prey; or it may remove its possessor from competition with the others of the same species by fitting it to another set of conditions. In any case, the struggle or pressure of life on the one hand, and on the other the fact that variation does occur and can be transmitted, will lead to what Darwin called *Natural Selection*-the survival of types possessing useful variations, in preference to any other types.

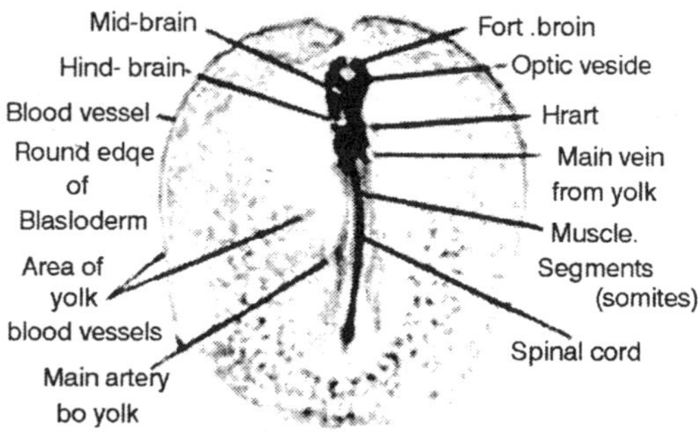

A

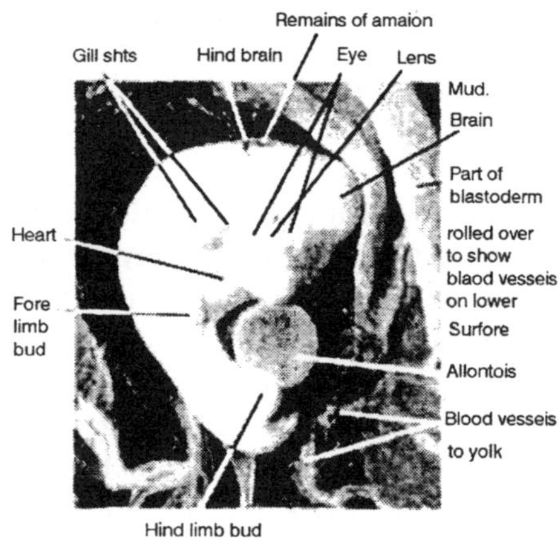

B

Fig. A And B. Embryo Chicks of About 36 Hours' and 4 days' incubation respectively

A has been stained and is photographed by transmitted light. In it about sixteen muscle-segments have been formed, the three main divisions of the brain are visible, with the eye-vesicles growing out from the fore-brain. The heart can be seen, together with a network of small blood-vessels over the yolk in the outer region of the blastoderm.

B has not been stained and is photographed as an opaque object by reflected light. The amnion has been removed (small traces of it are left in the head-region). The embryo rests on its left side, and the head has bent over. The limb-buds, gill-slits, and allantois can be seen. The blastoderm in one place has rolled over, showing the way the blood-vessels run on its lower surface, next to the yolk.

This will explain the fact that animals are adapted to their surroundings, since the unadapted or less well adapted could not survive; it will also explain the fact that when we find a number of species of animals all possessing the same general plan of structure, the different species are generally adapted to live in different conditions, since any variation enabling its possessor to live in new conditions will remove it from competition with the other members of the species. For instance, the general plan of all frogs and toads is very similar; but some live near water, others in drier places, some altogether in water, others on trees, and so on.

The only explanation, in fact, of the resemblance in structure between different frogs and toads is that they are all in very truth related, descended through millions of generations from a common original ancestor; and the explanation of their differences is that they are due to variation and natural selection, the random raw material of the former being sifted by the latter.

The backbone of the evolution theory is that all species have descended from other pre-existing species, and the backbone of the generally current explanation of evolution is that adaptation is due to variation (mutation) acted upon by natural selection.

Now frogs and toads resemble newts and salamanders in many points of their organization; for instance, in their moist, scaleless skin, in the structure of their heart, and in their development through a tadpole stage; and accordingly, they are all classed together as Amphibians. Snakes, lizards, tortoises, and crocodiles, on the other hand, have scales, their heart is a trifle more complicated, and they all develop inside a large-yolked egg with special protecting membranes round them; and they are accordingly all classed as Reptiles.

When we classify a number of animals together in this way, we are not merely pigeon-holing them for convenience; we pigeon-hole them in this one particular fashion because we believe them all to be descended from a common original ancestor. The classification aimed at by Zoology is a Natural Classification, aiming at a grouping of animals according to their blood-relationships. Thus, we believe that all frogs, toads, newts, and salamanders are descended from a common ancestral form with an Amphibian organization; and all snakes, lizards, and other reptiles from a common ancestor with a Reptilian organization.

But the organization of Amphibia has many points in common with Reptiles, as also with Fish, and with Birds and with Mammals, including, as we previously pointed out, man himself. All of these animals possess gill-slits and a notochord at some stage of their development, a backbone, red blood, a nervous system lying along the back, and so forth; they are therefore all classed as Vertebrates (or Chordates, with reference to the notochord), and there is no escape from the conclusion that, in the course of an exceedingly long space of time, to be reckoned in tens or perhaps hundreds of millions of years, they have all descended from an original common ancestor.

Belief in the occurrence of evolution, quite apart from any theory of how it has happened, is forced upon us by numerous sets of facts, which remain quite unexplained on any other supposition. In the first place, as already mentioned, there is the fact that animals can be arranged in groups (such as the Vertebrate group), each group with a single general plan of organization; and that within each group there are subgroups with their own special modification of the general plan (e. g.

the Birds or the Amphibians within the Vertebrates). Secondly, we have the fact that animals during development often pass through stages in which they resemble other and usually simpler animals of the same group. The frog in the tadpole stage has gills, and even the embryos of fowls and men possess gill-slits.

Both frog and man, although tailless when adult, possess tails when young. All this is very difficult, if not impossible, to explain unless we suppose that our remote ancestors, and those of birds and frogs as well, lived in water and possessed gill-slits and gills.

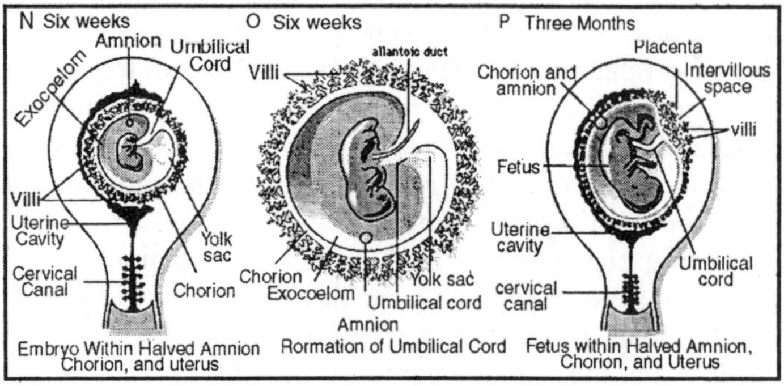

Fig. An Early Human Embryo

Thirdly, we have the existence of vestiges. Vestigial, or as they are often incorrectly called, rudimentary organs, are organs which are of little or no use to their possessor, although in related forms they are larger and obviously have a function. In man, for example, the hairs on the body, the fold of skin at the inner corner of the eye, and the skeleton of the tail are wholly or partly vestigial.

They exist to-day chiefly because our ancestors possessed them in more developed form. The fold of skin in the inner corner of our eyes, for instance, is the remains of the large movable third eyelid of many lower animals.

In whales the whole hind-limb has become vestigial. If you look at the hairs on the outer side of your arms, you will find that on the upper arm they point obliquely downwards, on the fore-arm obliquely upwards—all of them, in fact, towards the elbow. In members of the three classes Reptiles, Mammals, and Birds, efficient flying-organs have independently evolved (convergence). The fore-limb is always utilized as the main part of the wing, and its general plan is retained throughout (homology).

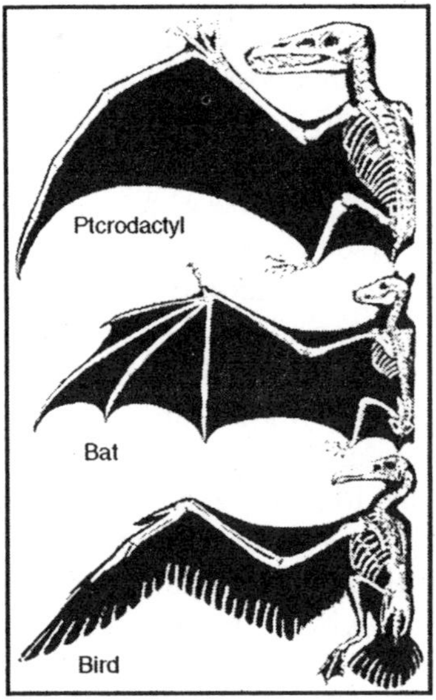

Fig. Homology and Convergence

But the details are different in each case. The main support (apart from the upper and lower arm-bones) is, in the Pterodactyl, the 5th or 'little' finger; in the Bat, the 2nd to 5th fingers; in the Bird, the quills of the feathers. Accordingly, only in the Bird is the hind-limb not required as part of the support of the wing, and is left free for other functions.

That is no good to you. But it appears that some of the apes have the habit of keeping their bodies dry in the rain by clasping their hands in front of their neck. The rain runs down the hairs towards the elbows; these are a little out from the sides, and the long hairs stick out from them to form a kind of spout, from whose end the rain pours off right away from the body.

Fourthly, we have the fact that in species of the same

group which are adapted to different methods of life, the same plan recurs in organs adapted to the most different uses. For instance, in the skeleton of the fore-limb in all land animals, although bones may fuse together in some, or altogether disappear in others, the same general plan recurs throughout. This one plan can be seen in the supporting limb of a frog, the paddle-like fore-limb of a whale, the running limb of a horse, the flying limbs of birds and bats, the arm of man.

Great difference in detail of adaptation, together with great similarity in general plan—it is difficult to account for this except by common ancestry and a common plan gradually modified in the course of evolution.

Finally, and most conclusively, the fossil remains of animals from earlier periods of the earth's history often show us actual intermediate stages in evolution. For example, the horse has in its forefoot but one finger which it uses, together with two tiny vestigial fingers, the splint-bones.

If we accept the idea of evolution, we must suppose that these vestigial fingers were once used; and as a matter of fact, if we look at fossils from a certain period we find no true horses, but animals which, though very like horses in most ways, possess three well-formed toes on the foot.

Again, to take an illustration on a large scale as well as one on a small, if we go back steadily in the history of the earth, we come to a time when man did not exist, or at least no traces whatever of his existence are to be found preserved in the earth's crust.

Then, long previously, to one when all the mammals were small and primitive, and all the birds toothed; then, to a time when there were no mammals or birds at all, but great reptiles, many of types now unknown, were the dominant living things; before that, to a time without reptiles, when amphibia were the only land vertebrates; and before that again, to a world without land vertebrates at all, but still with fish in the sea; and finally, a stage is found in which there are no fish, but only marine invertebrates.

The study of heredity showed us that each individual arose from actual portions of living substance which had once formed part of other individuals. The study of evolution shows that species arise from other species.

Within a single species there are a number of parallel streams of living substance flowing through the generations; but these parallel streams may diverge, and the original species branch into two.

Since each species has evolved out of another species, and each individual grows from a detached part of another individual, the whole of life must be looked on as a single mass of living substance, flowing on in time, and divided in the course of its history into a number of separate courses of different size, the main groups, the smaller groups, and the species. The alterations in the branches of this mass of living substance we call Evolution.

Chapter 3

Exchanges of Matter and Energy

Although there are many features of animal and human life which have not yet been, and perhaps never will be, explained in terms of physics and chemistry, still we can apply to animals, including man, certain great principles, such as the laws of the conservation of mass and energy, without restriction. It will be necessary to give a brief proof of this statement, for if it were not true we could, for example, never be certain that a man was not obtaining energy from unknown sources, in which case physics would be of very little use to the biologist.

If we put a man on a very sensitive balance it is impossible to obtain his weight quite accurately, because at every swing the scale containing the man rises slightly higher. He is losing weight, that is to say losing matter, at every moment of his life. Obviously some of this matter is water-vapour.

At the body-temperature the breath is saturated with watervapour which condenses into drops on a cold surface or in cold air. Besides this there is a constant slight loss of water-vapour from the skin even in the coldest weather, and in very hot weather a man may lose over a kilogram of sweat an hour. Besides water-vapour we are always losing carbon dioxide in our breath, and a very little from our skin.

This may be easily shown by breathing out through a tube dipped into lime water (CaO_2 H_2). The carbon dioxide of the breath combines with this to form a white precipitate of chalk ($CaCO_3$).

We can measure the production of CO_2 and H_2O very accurately by putting a man, or better an animal such as a dog or mouse, into an air-tight box. Air is passed into this box through bottles containing strong sulphuric acid (to absorb water-vapour) and caustic soda (to absorb carbon dioxide). On leaving the box it passes through similar bottles, and by weighing them we can determine the animal's output of carbon dioxide and of water.

Nothing else in weighable quantity has been added to the expired air, but it has lost oxygen, as may be proved by taking a sample and making the oxygen in it combine with hydrogen or some other easily oxidizable body.

The man or animal has therefore gained some oxygen from the air and given up carbon dioxide and water to it. Besides these, in the course of a day he will eat, drink, and excrete, and we can weigh his food, drink, and excretions, and thus construct a balance-sheet. The following represents a typical day's balance-sheet for a resting man weighing about ten stone.

Gain.	*Loss.*
Food, 1·1 kilograms (of which 0·60	Solids (mostly dissolved in urine),
kilogram is water).	70 grams.
Drink, 1·5 kilograms of water.	Water excreted, 1·3 kilograms.
Oxygen, 0·7 kilogram, or 500 litres.	Water evaporated, 1·1 kilograms.
	Carbon dioxide, 0·82 kilogram, or
	425 litres.

This leaves him with a gain of 10 grams for growth, and if he is weighed before and after the experiment he will be found to have gained 10 grams. It is clear from the above balance-sheet that much the most important chemical change taking place in the man is the combination of the carbon and hydrogen of his food with the 0·7 kilogram of oxygen to form 0·3 kilogram of water and 0·82 kilogram of carbon dioxide.

Now this is exactly the same change which occurs if we burn the food, a fact discovered by Lavoisier in the eighteenth century. Hence, if we put our man or animal into a calorimeter for a day and measure his heat-production, and then burn in another calorimeter a quantity of food exactly equal to the amount which he has eaten, we can make an energy balance-sheet like the mass balance-sheet produced above.

Some of the solids excreted are not entirely oxidized, so we must burn them in a calorimeter and subtract the energy thus obtained from that of the food. We must also allow for the energy of any food which is stored up but not oxidized. If the man is in a calorimeter all his external work will be converted into heat by friction; for example he may be made to ride a fixed bicycle with a brake.

Gain.	*Loss.*
From food, 2,400 kilocalories.	Heat lost by conduction and con-
Less waste from u oxidized excreta,	vection, 1,500 kiloc lories.
150 kilocalories.	Latent heat of water evaporated,
Net total gain, 2,250 kilocalories.	630 kilocalories.
	Heat used in warming food and
	drink to body temperature, 60
	kilocalories.
	Total loss, 2,190 kilocalories.

This leaves him with 60 kilocalories stored up (probably mostly as fat) for future use. It has been found by experiment that the more accurately all the measurements necessary are made, in an experiment of this kind, the more nearly do the gain and loss of energy balance, so long as the body neither gains nor loses matter.

That is to say, the law of the conservation of energy holds for men and animals. They do not obtain energy from any mysterious sources, nor do they convert it into any forms which cannot be reconverted into heat. The energy arises from the recombination of oxygen with carbon and hydrogen, from which it has been separated by plants.

The carbohydrates, fats, and proteins formed by the plant are largely eaten by animals, which use them partly for growth and repair, but mainly as a source of energy. A carnivorous animal obtains its energy secondor third-hand from the plant, but in the long run all animal energy, just like the energy derived from coal, is stored sunlight. We can now consider the energy values of different foods, and the energy requirements of different classes of people.

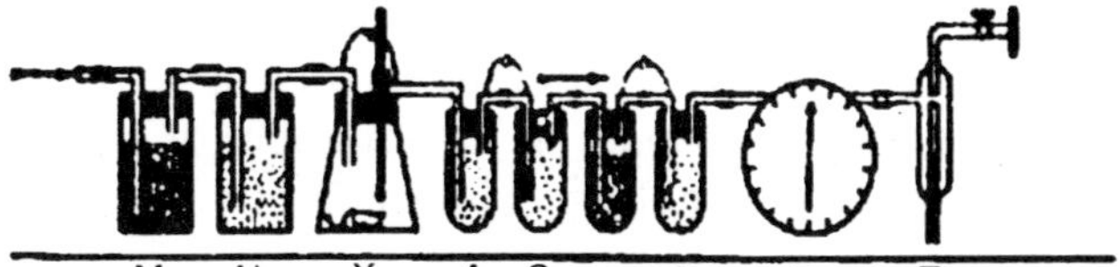

Fig. Method of Determining the Gaseous Exchange of a Small Animal. N, A, B, and D Contain pumice soaked in strong sulphuric acid to absorb water vapour. X contains a mouse, M and B a mixture of soda and lime to absorb carbon dioxide. Air is sucked through by the pump (Z) on the right, to which is attached a gauge (Y). The mouse therefore gets air free from CO_2 and H_2O. The H_2O breathed out by it is caught in A and B, its CO_2 in C, and any H_2O evaporated from C in D. Hence the increased weight of C and D gives the animal's CO_2 output, the increased weight of X, A, B, C, and D together its O_2 consumption.

Foods consist of water, salts, carbohydrates, fats, and proteins, of all of which some account has been given in former volumes of this series. Energy can be obtained from the oxidation of the latter three. Proteins are not completely oxidized by animals, so we have to deduct the energy value of the unoxidized remnant from that of the original protein. When this deduction has been made we find that the energy values are as follows:

1 gram carbohydrate	=	4·1	kilocalories
1 gram fat	=	9·3	"
1 gram protein	=	4·1	"

That is to say, fat is more than twice as good a source of energy, weight for weight, as any other food. It must be remembered that while fatty foods such as butter and lard contain very little water, and bread, which is our greatest source of carbohydrate, contains only about 10 per cent.; lean meat, though most of its solids are protein, contains 75 per cent. of water. We obtain most of the energy we need from carbohydrates and fats. The proteins, though they also supply energy, are mainly required for the growth and repair of our bodies. The accompanying table shows the energy values per pound of a number of foods, and the number of calories that could be bought for a shilling in July 1926.

	Kilocalories	*Kilocalories*	*Protein*
Food	*per lb*	*per shilling*	*per cent*
White Bread	1,200	5,760	8
Biscuits	1,900	1,750	10
Oatmeal	1,810	5,430	15
Sugar	1,815	6,400	0
Milk	314	1,260	3·3
Butter	3,490	2,090	1
Cheese (Cheddar)	2,080	2,500	28
Lard	4,090	4,260	0
Beef (round, lean)	650	650	20
Beef (rump, fat)	1,360	630	13
Bacon	2,370	1,580	9
Herrings (fresh)	360	720	11
Potatoes	300	3,610	1·8
Cabbage	120	1,450	1·4
Apples	215	430	0·3

The energy requirements of a man or animal vary, like those of a machine, according to the amount of work he is doing, but, unlike a machine, an animal, even when resting completely, needs a considerable amount of energy, partly for internal movements, such as those of the heart, the gut, and the muscles concerned in breathing, partly, in the case of warm-blooded animals, to maintain the body-temperature.

Above this minimum, additional energy is required according to the amount of work done. An individual muscle, say the biceps of the upper arm which bends the elbow, may have an efficiency as high as 40 per cent., that is to say it can turn into work nearly half the energy that it derives from food and oxygen, but if we consider the body as a whole the efficiency is a good deal less, since the heart has to work to supply the muscle with its blood, the lungs and gut to supply the blood with oxygen and food, and so on, in which process much potential energy is converted into heat by internal friction. As a matter of fact, we find that the efficiency of the body as a whole is never more than about 25 per cent.

Even this, however, is greater than the efficiency of any heat-engine of the same weight. The advantage of mechanical over animal power is not that the machine is more efficient, but that its fuel is cheaper, and that it does not waste energy while it is not working.

If we take a man's output of energy per minute when lying down as 1, it will be about 1·4 when he is standing, 4 when he is walking at three miles an hour or doing moderate work with his arms, 10 during fairly hard work, 15 during the most violent exertion of which he is capable for any length of time, and over 50 during very violent work, such as a hundred-yard sprint.

His intake of oxygen and output of carbon dioxide vary directly with the amount of energy set free, and if he is to keep up his weight he must eat enough food per day to supply the energy which he is liberating. The following table shows the energy requirements in kilocalories per day of different workers each weighing about 65 kilograms (10 stone).

Profession.	*Energy needed in food.*
Clerk	2,410
Tailor	2,510
Cobbler	2,940
Metal-worker	3,290
House-painter	3,500
Carpenter	3,550
Stonemason	4,660
Woodcutter	5,400
Cyclist (racing for 16 hours)	10,240

The cyclist probably could not eat the required amount of food per day, and so had to use up his own fat as a source of energy. The energy output of a resting warm-blooded animal (mammal or bird) is proportional to its surface, not to its weight or volume. Thus if two dogs are of the same shape, but one twice as long, high, and broad as the other, it weighs eight times as much, but needs only four times as much food per day, as its surface is only four times that of the small dog. Thus children need a great deal more food per pound of their own weight than do adults for energy-production alone, besides their requirements for growth.

Similarly if we compare a large cart-horse weighing a ton with a ton of men (13), a ton of fowls (500), and a ton of mice (55,000), their food requirements will be proportional to their total surfaces, and the men will need 3·7 times as much energy per day when at rest as the horse, the fowls 8 times, and the mice 25 times. This law does not hold for 'cold-blooded' animals.

Although in the long run the oxidation of food is our only source of energy, yet a muscle or gland can work for a short time without any oxidation. The immediate source of muscular energy seems to be the breakdown of a compound of sugar and phosphoric acid into lactic and phosphoric acids. This chemical reaction is used as a source of work, but to put its products together again so that they can furnish more work, sugar must be oxidized, so a muscle cannot work for long without being supplied with oxygen.

In a sprint of 100 yards the leg-muscles work faster than we can supply them with oxygen, and are very short of it at the end. So if in running a quarter-mile, a runner sprinted during the first 100 yards, his muscles would be short of oxygen for the rest of the race, and would work inefficiently. It is better to put off the sprint to the end and run less rapidly at first, so that the muscles are able to get all the oxygen they need.

Our energy requirements can be made up in various ways, just as a fire may burn coal, wood, or peat. Thus a diet of 135 grams protein, 80 grams fat, 500 grams carbohydrate will furnish 3,350 kilocalories, but within fairly wide limits we can replace one food by another yielding the same amount of energy. Fifty grams of the protein could be replaced by 50 of carbohydrate or 21 of fat without reducing the protein intake below what is needed for repairs.

But we cannot make out a dietary on a basis of energy alone. The tissue-proteins are always breaking down, and need a certain minimum of protein food to build themselves up again; whilst a growing child, a pregnant or nursing woman, or a man recovering from illness or building new muscle during training, needs more than this. All kinds of protein are not of equal value.

A vegetarian, if he does not take milk or cheese, must eat more protein than a meat-eater to keep in health, as vegetable proteins are generally less digestible than those of animals, and on digestion yield products which are not in the best proportions for building animal tissues.

We also need various inorganic substances besides water, for example, calcium salts for our bones, iron salts for our haemoglobin, sodium chloride for our sweat. Finally, our diet must contain small quantities of at least four, and probably five or six, different organic compounds generally called vitamins. We do not know the exact composition of these, though we know a good deal about them, for example, that one is a base (like quinine or adrenaline), two others soluble in oil like a fat or wax, and so on.

Vitamin A is a fat-like and oil-soluble substance found in leaves and many natural fats and oils. It is necessary for growth, and a shortage of it leads to eye troubles and loss of immunity to various diseases. *B* is easily soluble in water and probably a base. It is found in many living tissues, notably yeast and the embryos of cereal grains. Its absence leads to a failure of growth, and to affections of the nervous system.

C is water-soluble and easily destroyed by oxidation. It is found especially in fruit and leaves. Its absence leads to scurvy. *D* has properties and a distribution similar to those of *A*. Its absence leads to rickets and bad teeth, but it is formed whenever an at present unknown precursor, related to cholesterol, and a constituent of most cells, is acted on by ultra-violet radiation.

So a plentiful exposure of the skin to sunlight will make up for a deficiency of it. *E* is an oil-soluble substance found especially in wheat embryo.

Its absence causes sterility, but not ill-health. This list is probably not complete. Some of the vitamins, especially C, are liable to be destroyed by over-cooking, others are often removed with the bran in milling grain, so that human diets are often short of one or another of them. Most plants can make some or all of them, and some animals do not require them all.

To sum up, a complete diet must include inorganic substances, organic bodies which can be oxidized to yield energy, and a number, probably about twenty, of different organic compounds which the animal body cannot itself build up.

Chapter 4

Transport in the Body

Energy is constantly being liberated in every part of the body as the result of chemical changes, so we must study the methods by which food and oxygen are distributed, and waste products removed. We may begin with the exchange of gases, which in our own case we call breathing. In simple animals like jellyfish and many worms there are no special organs for this purpose.

Dissolved oxygen soaks through their skins from the water in which they live, and dissolved carbon dioxide soaks out. In somewhat more complicated creatures like the earthworm there is a special fluid, the blood, one of whose functions is to carry oxygen from the skin to the internal organs, and carbon dioxide back. In the most advanced water-dwellers we find special tufts of thin skin, the gills, into which blood is pumped by a heart or hearts, so as to expose it to water with which it may exchange gases.

Gills may be naked, as in the young tadpole, but usually, as they are very delicate, they are protected by lids or pouches, as in fish or lobsters, where they lie behind the head on each side. In air-breathing animals there are two quite different kinds of breathing organ. In insects and spiders, air is carried to every part of the body by tiny branching tubes called tracheae, which open by numerous pores, mainly on the sides of the abdomen.

In air-breathing vertebrates and molluscs the blood is exposed in the lungs to air which is continually renewed by the act of breathing. Some animals combine several methods. Thus a frog can use its skin alone for breathing so long as it does not need much oxygen, but it normally employs its lungs as well. The human lungs are elastic organs of a spongy texture, consisting of millions of very small cavities which open into stiff tubes called bronchi. These in turn open into the windpipe in the front of the neck. Since air cannot enter the space between the lung and the chest-wall, the lungs expand when the chest is expanded.

Fig. Final Ramifications of Tracheae in a Caterpillar

This can happen in two different ways. In the first place the diaphragm, a sheet of muscle separating the chest and belly, and bulging upwards into the former, may contract and force the contents of the belly downwards and outwards. This pulls the bases of the lungs down, and draws air into them. Or the muscles which lie between the ribs may contract so as to bring the ribs, which normally slope downwards, into a more horizontal position.

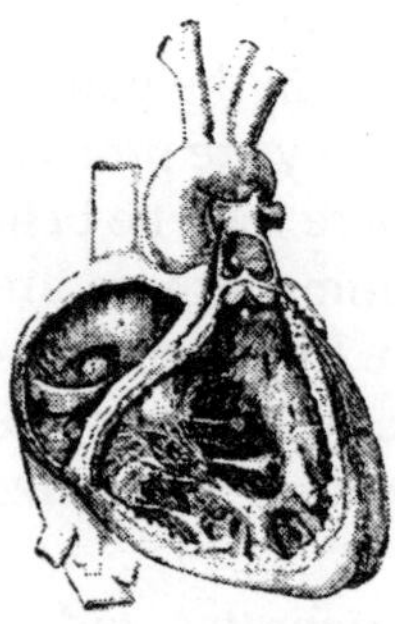

Fig. A Human Heart

The breast-bone is thus pushed away from the vertebral

column, while at the same time the diameter of the chest from side to side is increased, and the lungs expand to fill the extra volume. Muscles acting in the opposite direction force the air out again if necessary; generally the elasticity of the lungs is sufficient. Thus the air to which the blood in the lungs is exposed is constantly being changed. The inspired air contains 21 per cent. of O_2 and only 0.03 per cent. of CO_2, whilst the expired air (after being dried) contains about 17 per cent. of O_2 and 3 per cent. of CO_2. To understand how this change occurs we must study the circulation of the blood.

The blood flows to and from the heart through a closed system of tubes known as arteries, veins, and capillaries. It leaves the heart by the arteries, which are comparatively thick-walled, and seldom near the surface, and returns by the veins, which have thinner walls, and often lie just below the skin. It passes from the arteries to the veins by the capillaries, which are too small to be seen with the naked eye, but are found in almost every tissue.

The most easily felt arteries in man are those of the wrist and temple, the most easily seen veins those on the back of the hand and foot. Finally, the heart, which pumps the blood round, is a hollow muscle with four chambers, lying between the lungs, and about as large as the two fists together.

The blood from all parts of the body except the lungs enters the heart from behind, near the right-hand top corner, and flows into a chamber called the right auricle. This contracts rhythmically (the average rate in a grown man at rest is about seventy times a minute) and forces its contents into a thicker-walled chamber lying below it, the right ventricle. As soon as this is full it contracts in its turn.

The blood is prevented from returning into the auricle by a valve, called the tricuspid. It therefore finds its way out through the pulmonary artery, which leads it to the lungs. The semilunar valves at the base of the artery prevent it from flowing back into the ventricle when this relaxes at the end of its stroke. In the lungs it has to pass through capillaries in the walls of the air-sacs, so that it is only separated from the air by a very fine membrane, and can easily lose its carbon dioxide and take up oxygen. It returns from the lungs by the pulmonary vein, this time to the left auricle, which contracts at the same time as the right auricle, and fills the left ventricle. The left ventricle, which is assisted by valves not unlike those on the right side of the heart, forces the blood into the largest artery, an elastic tube called the aorta, from which it is distributed all over the body by the other arteries.

One way of forcing it round through the narrow capillaries would be to have a cistern at the top of the head from which it flowed down again, but this arrangement would clearly not work when its owner lay down. In reality, a fairly steady pressure is kept up by the fact that the walls of the aorta and the arteries are always stretched, and continue to squeeze the blood along between the strokes of the ventricles.

The pressure in man is about that of a column of blood 1.7 metres high, i.e. equal to the head of blood which would be obtained from a cistern 1.7 metres above the heart. The heart of a grown man at rest delivers about 8 litres per minute, or just over 100 cubic centimetres per beat. During exercise the heart may deliver three or four times as much per minute, mostly as the result of an increased number of beats per minute. The beat of the heart which is felt below the fifth rib on the left side is due to the left ventricle striking the wall of the chest each time it contracts and stiffens.

The blood is squirted along the arteries at a rate which may be as high as 50 centimetres a second, and each fresh wave causes a pulse in the artery. If an artery is cut, the blood comes in a series of spurts from the side nearest the heart, and can be stopped by pressure on the heart side of the cut. It is easy (and safe) to stop the pulse in the wrist by pressing on the same artery higher up, either inside the elbow or inside the upper arm just below the armpit, where it can be felt.

The blood flows through the capillaries very slowly, at about half a millimetre per second. Their average length is about a millimetre, while their diameter may be less than 0.01 millimetre. Their very thin walls allow water, gases, and dissolved substances to be exchanged between the blood and the tissues with great ease. From the capillaries the blood oozes gently into the veins. They have no pulse, and a comparatively small pressure and rate of flow. A cut vein bleeds steadily and the flow can be stopped by compressing the side away from the heart. The flow of blood in the veins is assisted by the presence of valves in them, which only allow it to move towards the heart. If the finger-tip be run along one of the veins of the forearm away from the heart, the vein will dilate on the heart side of each valve. When therefore the contraction of various muscles squeezes the veins, the blood can only flow towards the heart. If a man stands quite still the blood tends to accumulate in the veins of his legs, and he is liable to faint from failure of the supply to his brain. When he starts walking the blood is at once squeezed out of these veins.

The only veins in man which do not lead straight back to the heart are those from the digestive canal and some other abdominal organs, which pass into the liver. Here the food absorbed by the blood from the gut is dealt with. As the liver needs oxygen, it has also a supply of fresh arterial blood.

In most other animals the circulation is somewhat different. Thus in a fish the blood only goes once through the heart in a complete journey, instead of twice, and it all passes through the gills before going on to the tissues. In the frog, in keeping with its two ways of breathing, we have a condition almost half-way between that found in fish and men.

Fig. Diagram of the Course of the Circulation in Man

The blood entering the right side of the heart by the main vein flows into *R.A.*, the right auricle; thence, through a valve, into *R.V.*, the right ventricle, which pumps it, through a valve, down *P.A.*, the pulmonary artery, to the lungs, *Lg*. The blood here flows in capillaries, and is then gathered up into the pulmonary vein, *P.V.*, and taken to the left side of the heart, entering the left auricle, *L.A.*, then (through a valve) the left ventricle, *L.V.*, and by this being pumped out (again through a valve) into the aorta, *Ao*.

Some of the blood (A^1) goes to the head and anterior extremities, whence it is returned by the anterior venae cavae (V^1), into which discharges the main trunk of the lymphatic system, the thoracic duct (*Th.D.*). The rest (A^2) goes to the trunk and hind limbs. That which supplies the digestive tube (*Al.*) passes to the liver (*Lr.*) by the hepatic portal vein (*V.P.*); the liver also receives arterial blood direct by *H.A.*, the hepatic artery.

The hepatic vein, *H.V.*, joins the veins from the other posterior regions, V^2, and flows to the heart in the inferior vena cava, *V.C.I. Lct.* denotes the lacteal lymphatic vessels from the intestinal wall, *Ly.* ordinary lymphatics.

We must now consider how the blood acts as a carrier of oxygen and carbon dioxide. Water at body-temperature will only take up one two-hundredth of its volume of oxygen from air; if the blood were no better than this we should need a heart working forty times as fast as the one we have. Actually the oxygen is carried round in loose chemical combination with a body called haemoglobin.

This is a protein, but contains iron as well as the usual carbon, hydrogen, oxygen, nitrogen, and sulphur. It is of a purple colour, but becomes red on combining with oxygen, which it readily takes up from the air. Thus the blood in the veins is purplish, but if exposed to air either in the lungs or by opening a vein, it at once becomes red.

The haemoglobin which it contains enables blood to hold 18 volumes of oxygen per cent., which is 6/7; of the amount contained in the same volume of air, and about 40 times what the blood could carry without haemoglobin. Since blood carries dissolved solids as well as gases, a single set of capillaries supplies the tissues with all that they need for activity, growth, and repair, besides removing most of the waste products. The carbon dioxide produced in the tissues is mainly carried, not in solution as such (CO_2), but in combination as sodium bicarbonate ($NaHCO_3$).

If we look at a drop of blood under a microscope we see that it consists of a clear fluid full of little reddish-yellow biscuit-shaped bodies about 0.007 millimetre in diameter. They can just be seen with a powerful hand-lens. A cubic millimetre of blood contains about five million, and as a man has about 4 litres of blood the total number in his body is about 20 million million, or far more than the number of men who have lived since history began, or probably at all.

Their total surface is about 1,500 times the surface of the body. All the haemoglobin of the blood is contained in them, and they therefore carry round the oxygen from the tissues. Their huge surface area renders this exchange easy.

They are cells which have lost their nucleus and are therefore not fully alive, but act as passive carriers of oxygen. They are always being produced in large numbers in the bone-marrow, and destroyed in the spleen and liver when worn out. For every 500 or so of these red corpuscles there is one white corpuscle.

Fig. A Large Capillary Vessel

They are true cells with nuclei, and some of them are capable of active movements. There are at least six different kinds with different functions, mostly of a protective character. Thus one kind eat up parasites found in the blood, another kind burrow through the walls of capillaries in inflamed areas, and remove dead or injured tissue and disease-germs, often forming collections of 'matter' or pus. Others produce substances which kill disease-germs, and so on.

Fig. A Small Arteriole (Left) and Venule

The blood also contains non-cellular bodies called platelets, which are smaller than the corpuscles and are concerned in clotting and in immunity; there may also be tiny drops of oil after a fatty meal.

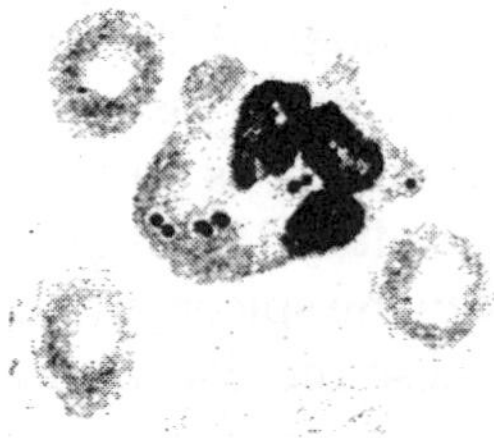

(*a*) are seen one white and three Red Blood-corpuscles. The White Corpuscle has a large Nucleus in three parts, and has ingested a number of *Micrococcus Pyogenes Aureus*, the common Bacterium of Boils, Abscesses, and c.

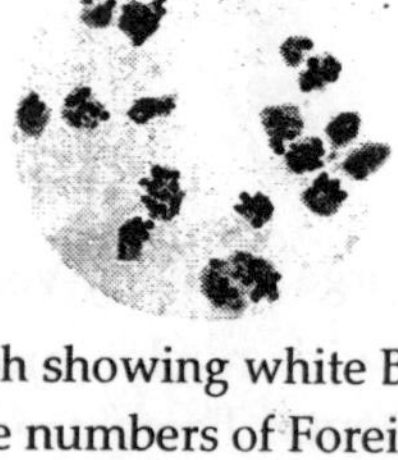

(*b*) is a Microphotograph showing white Blood-Corpuscles which have ingested large numbers of Foreign Particles (Sheep Red Blood-Corpuscles) with which they have been Incubated.

Fig. Two Examples of Phagocytosis by Human White Blood-cells.

It is seen to be connected with other lymphatic trunks and glands (e.g. those in the lumbar region, *d*), and to open into the junction of the left jugular (*f*) and subclavian (*g*) veins at *c*. It lies just ventral to the spinal column, on either side of which are seen portions of the ribs (1). *h*, anterior vena cava, just anterior to which, in the median line, is a piece of the wind-pipe.

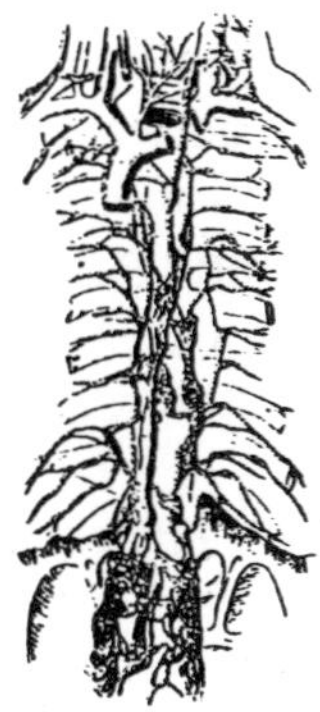

Fig. Ventral View of the Human Thoracic Duct

Besides the blood-vessels, all vertebrates possess another system of vessels which open into the veins, but contain clear fluid called lymph. The lymphatic vessels slowly drain away fluid from the spaces between the cells. On the course of the vessels the fluid passes through small lymph-nodes or glands, which may be felt under the skin in the neck, armpit, or groin, before entering the blood.

The lymph contains white corpuscles which are largely produced in these glands, and abnormal bodies from the tissues are dealt with in them. Thus if the arm is inflamed, the lymph-nodes of the armpit are generally enlarged, as they are busy destroying poisons or bacteria from the inflamed tissue. The lymphatics also play a part in digestion, which we shall study later.

The cells obtain the food-stuffs which they require. In simple animals such as polyps every cell is so close to the digestive cavity that it can obtain food thence directly. In most higher animals, however, the cells which line the alimentary canal and its glands are highly specialized for the purpose of breaking up the food-stuffs into soluble forms and passing them rapidly into the blood or other body-fluids.

This process is called digestion. A few internal parasites like the tapeworm have no gut, but absorb their food through their skin, relying on their host to break it up for them. The course of digestion in man is as follows. The food is chewed in the mouth, where it is also moistened by saliva, a fluid secreted mainly by three neighbouring pairs of glands.

Two pairs of these lie between the tongue and the lower jaw, the third pair (whose inflammation causes mumps) lie below the external ear, mainly inside the lower jaw-bone. Besides moistening the food, saliva contains an enzyme, ptyalin, which breaks up starch into an easily soluble sugar called maltose.

Enzymes play a much larger part in digestion than do mechanical processes. Each digestive enzyme is a definite substance with the property of bringing about, or enormously speeding up, a particular chemical reaction. Pepsin from the stomach will split up half a million times its weight of protein, but will not alter starch or fat. So delicately is an enzyme adjusted to the substrate on which it acts, that as most food molecules are asymmetrical (as we know from their rotation of the plane of polarized light and their asymmetrical crystals) so are the enzymes that act on them. Enzymes have been compared to keys which open certain locks only. But they are like 'Yale' rather than ordinary keys.

For we can make in the laboratory a sugar or peptide (part of a protein molecule) which only differs from the natural variety in that its molecules are related to the natural molecules as a left hand to a right or an object to its image in a mirror. Enzymes will not act on these artificial substances, though they digest their mirror-images which occur in nature. So Alice, who went through the looking-glass in the story, could not have digested the looking-glass proteins and carbohydrates. She would have had to get her energy from fat and alcohol, whose molecules are symmetrical, and would finally have died of protein starvation.

Enzymes are not alive, and can still work when removed

from the body; but they are generally destroyed by boiling. The action of ptyalin can easily be shown. A solution of boiled starch gives a blue colour with iodine. If saliva is allowed to act on it for a few minutes at body-temperature or a few hours in the cold, the starch is broken down to sugar and loses this property. If the saliva is first boiled no change occurs.

When the food is swallowed, it is passed into the gullet or oesophagus. To get there it must pass over the mouth of the wind-pipe, and when any falls down this we choke. To prevent this, the breathing is stopped while we swallow, and the cartilages at the top of the wind-pipe which are concerned in voice-production are brought together so as to close the top of the larynx.

It is further protected by a cartilaginous lid at the back of the tongue called the epiglottis, which is pressed backwards over it by the food during swallowing.

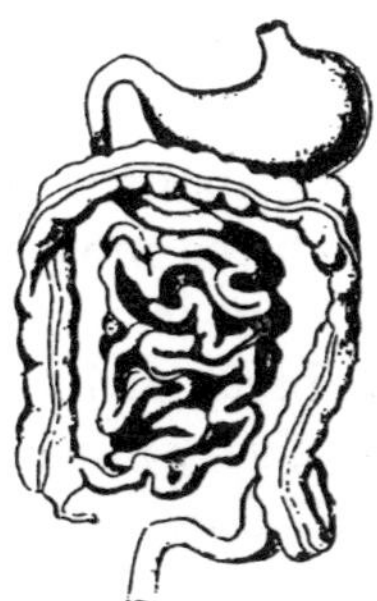

Fig. The Abdominal Portions of the Human Digestive Tube, Ventral View

The movements of the laryngeal cartilages can easily be felt. As the food leaves the mouth it passes out of the control of consciousness and will. The movements of our digestive canal, except at the two ends, are carried out by smooth or involuntary muscle and are controlled by a special part of the nervous system over which the mind does not preside.

The food or drink on leaving the mouth is seized by the smooth muscle of the oesophagus, which contracts behind it and relaxes in front, thus passing it rapidly down into the stomach. Owing to this gripping action one can eat or drink while standing on one's head. The narrow duct opens above into the cavity of the stomach, among columnar cells. The main part of the gland is a simple tube, formed of a single layer of cells; these secrete pepsin. A few darker-stained cells are seen near the outer side of the tube; these probably secrete hydrochloric acid.

Fig. A Simple Gland From The Stomach of a Mammal (A Bat)

The human stomach is a bag of smooth muscle lined with a membrane consisting mainly of microscopic glands. When expanded it will generally hold about 2 litres. The gastric juice is a clear fluid containing about ½ per cent. free HCl, and several enzymes, of which the most important is pepsin. In presence of acid (though not in a neutral or alkaline fluid), pepsin causes proteins to break up into bodies called peptones and proteoses, which have very much smaller molecules than the proteins from which they are derived, and are more soluble.

Some carbohydrates are attacked by the hydrochloric acid. Thus each molecule of cane-sugar is split into one of glucose and one of fructose, and inulin, a starch-like body found in many plants, is broken up into fructose. Fats are less affected in the stomach. A meal remains in the stomach for a time—generally between one and four hours—which depends on the nature and quantity of the food taken. All this time the muscular walls of the organ are contracting in such a way as to mix the food thoroughly with the gastric juice.

Hardly any absorption occurs in the stomach. When its contents are fully mixed the muscular ring surrounding the lower orifice of the stomach, the pylorus, opens, and a jet of its contents is squirted into the duodenum, the top portion of the small intestine. After a heavy meal the stomach may take several hours in emptying itself.

The human small intestine is a tube about 6 metres long and 3 cm. in diameter when relaxed. It is lined by fine projections of the mucous membrane, the villi, which give it a velvety texture, and increase tenfold the surface available for absorption. Between the bases of the villi are the mouths of numerous microscopic glands, and just beyond the pylorus open the ducts of two large glands, the liver and pancreas.

The liver, among its other functions, secretes bile. This contains substances which have the property, possessed in a less degree by soap, of lowering the surface-tension of water in which they are dissolved. Drops of oil or melted fat tend to join up so as to have as small a surface as possible.

This tendency of the surface to shrink is prevented by the bile in the intestine, so when fat-drops are broken up there they do not coalesce again, but form a milky emulsion. Hence the fat-splitting enzyme made in the pancreas can act on a vastly greater surface of fat than would otherwise be available.

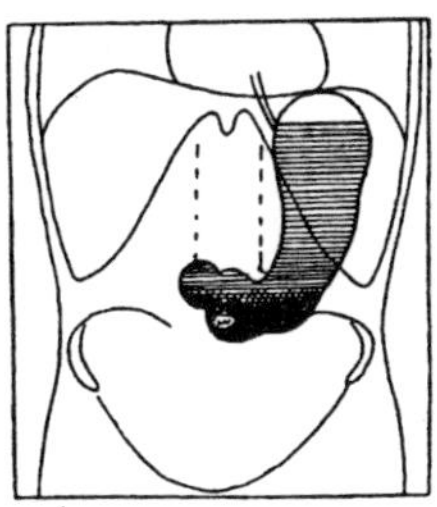

Fig. The position and shape of the moderately full human Stomach, as revealed by X-ray Photography after a meal mixed with oxychloride of bismuth, which is Opaque to X-rays. O, Gullet (Oesophagus); F, Fundus of the Stomach (containing air); PC, pylorus of Stomach, opening into the small Intestine; U, position of the Umbilicus (Navel) on the surface of the body. The dotted lines indicate the position of the backbone.

A man with jaundice from a blocked bile-duct can digest milk, whose fat is already broken up, but not suet or butter, which form big drops. The bile also contains pigments formed from the haemoglobin of worn-out red corpuscles. These are excreted in the faeces and give them their yellow colour. In jaundice the bile-duct is blocked, and accordingly the skin becomes yellow and the faeces white. Bile is stored in the gall-bladder till required.

The pancreas secretes a juice containing a number of enzymes. One of these, like ptyalin, breaks down starch into maltose. Others break down maltose to glucose, and lactose (milk-sugar) into glucose and galactose, a very similar substance. Another breaks down fat into glycerol and fatty acids, which partly combine with alkali to form soap. In every case the molecules formed will pass more easily through a membrane than those of the food.

If the juice contained a protein-splitting enzyme it would probably digest the pancreas itself and its duct. However, it contains a substance which, on mixing with the secretions of the intestinal glands, yields trypsin, an enzyme which attacks proteins and the products of peptic digestion, breaking them down into amino-acids. These enzymes will not act in an acid fluid, so the bile and pancreatic and intestinal juices contain enough sodium bicarbonate to neutralize the acid of the gastric juice.

The secretion of the intestinal glands, besides the substance which helps to form trypsin, contains an enzyme which will break down peptones, though not whole protein molecules, into amino-acids. There are also enzymes which break down milk-sugar and cane-sugar to sugars containing only six carbon atoms.

So far as we know, no animal more complicated than a snail produces an enzyme which will break up cellulose. But grass consists very largely of cellulose, and to digest it hoofed animals and other plant-eaters employ bacteria which grow in their digestive canal. In cud-chewers like the cow and sheep these bacteria live in special compartments of the stomach; in the horse in the large intestine, which may have a capacity of 200 litres.

The bacteria can get very little oxygen, so they cannot oxidize the cellulose, but they turn a good deal of it into methane, which is wasted, and leave much undigested. The rest, however, is broken up into small molecules which the animal can absorb. In man, cellulose is not digested, but it is useful in giving bulk to the faeces and preventing constipation, which easily occurs when the food leaves no indigestible residue. The food is thus broken up into easily soluble constituents, and is ready to be absorbed.

This is done by the epithelium of the small intestine. The passage through is not a mere filtration. For example, the blood contains one part of sugar per thousand, and if sugar merely filtered across from gut to blood, it could never quite disappear from the gut as it in fact does. Actually, during the absorption of food, the absorptive cells perform work, for which they need an extra supply of oxygen.

The fats are, in part at least, put together again from glycerine and soap, and passed as a fine milky emulsion into the lacteals, as the lymphatics of the small gut are called. These join together to form the thoracic duct, which runs up through the chest and empties into the jugular vein at the base of the neck. The blood may be noticeably milky after a heavy meal of fat. The rest of the products of digestion are passed into the capillaries, and dissolved in the blood.

All the veins from the gut run to the liver, and here the food is further dealt with. Thus sugar, if not needed immediately for oxidation, is stored as a starchy body called glycogen, discovered by Claude Bernard, and gradually liberated as required later on.

Some of the sugar and most of the fat are stored elsewhere. Part of the sugar is stored as glycogen in muscles, but much of it is made into fat and stored under the skin and round some of the internal organs, along with fat from the food. In the frog fat is stored in special fat-bodies. The stored glycogen is later split up into glucose for use in the body by an enzyme, which continues to work after death, and the sugar thus liberated gives liver its well-known sweet taste.

Absolutely fresh liver is not sweet. Again, ammonia, which is formed in the digestion of many proteins, is, in the liver, mostly combined with carbon dioxide to form urea (2

$NH_3 + CO_2 = CON_2H_4 + H_2O$). Urea is a very innocuous body, but ammonia rather poisonous if it gets to the brain, so that if the blood from the gut is short-circuited into the vena cava instead of going into the liver, a heavy meal of meat may cause convulsions. The liver also deals on the same sort of lines with any excess of amino-acids in the blood and with various poisons; its most important function is thus the regulation of the blood's composition, and not the secretion of bile. During starvation the body lives on its stores of fat, and the liver takes on the new duty of converting the stable fats such as those of suet into oils like that of linseed, which are very easily oxidized. The unabsorbed residue of the food from the small intestine passes into the large intestine, where it remains in man for a day or so, and is acted on by bacteria. These are of little or no value to man, though very valuable to some animals. In man little but water is absorbed there.

By this removal of water the bulk of the waste food is reduced by about nine-tenths. The large intestine also excretes poorly soluble salts, such as calcium phosphate, from the blood. These would clog up the urinary passages if they were excreted by the kidneys.

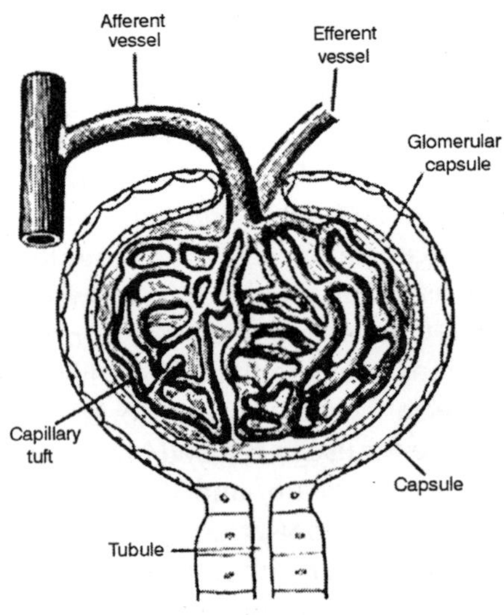

Fig. Diagram of A Malpighian Corpuscle in the Human Kidney

The active tissues take oxygen, sugar, fat, and aminoacids from the blood, and use them for oxidation, growth, and repair. Into the blood they empty waste-products, of which the most important are water, carbon dioxide, and urea, but other soluble waste-products of protein metabolism include sulphuric and phosphoric acids, creatinine, and uric acid.

The last two contain C, H, O, and N. They are all excreted by the kidney except the carbon dioxide and some of the water, which go out by the lungs, and, in the case of water, the skin. The kidneys consist of a mass of tubules (about a million on each side in man) each beginning in a capsule containing a tuft of capillaries, and ending, after a winding course, in the central cavity of the kidney.

The end of the kidney tubule is thin-walled, and is dilated to form a capsule (Bowman's Capsule). This is invaginated by the ingrowth of a small arteriole and venule, which break up within the inner wall of the invaginated capsule to form a network of capillaries or glomerulus.

The capillary tuft seems to act as a filter, and the fluid that soaks through it is blood minus corpuscles and minus a few of the large molecules, such as the proteins concerned in clotting. As this filtrate runs down the tubules, the cells lining them reabsorb valuable constituents of the blood, such as sugar, and probably add unwanted ones such as urea, ammonia, uric acid, creatinine, and sulphates. The urine trickles down the ureters into the bladder, whence it is emptied from time to time. An adult man produces on an average about 1.5 litres of urine per day, containing 30 grams of urea, 15 of sodium chloride, and 10 of other soluble waste-products.

The blood is thus the medium of exchange between the different parts of the body. The heart keeps it moving, the lungs and gut supply it with fresh oxygen and foodstuffs, other organs get rid of waste products, and all the tissues of the body take from it according to their needs. The life of every part of the organism therefore depends on an adequate supply of blood of proper composition.

Chapter 5

The Nervous System

The activities so far described may almost all be found in machines, but a complicated machine needs human control if the parts are to work harmoniously. The body to a large extent runs itself. The diaphragm and heart contract with the appropriate force and rhythm, the pancreas begins to secrete as food reaches the duodenum, and so on.

But in many of our more complex activities, such as the movements of my hand consciousness plays a part. The chief agency in co-ordinating our actions is the nervous system. Many of its activities are unconscious: consciousness and will only play a part where our past experience is likely to be of value in influencing our behaviour.

The unconscious responses of the nervous system are called reflex actions, or reflexes; though of course it is also responsible for voluntary actions. The nervous system controls striped muscle, heart muscle, smooth muscle, and glands, but with very few exceptions it is only the striped muscles that the will can influence, and even they are often moved by reflexes. In every reflex or voluntary action three organs are always concerned, first a receptor organ which is appropriately stimulated, then a longer or shorter path in the nervous system, and finally an effector organ.

The latter is always a muscle or gland in man, though other animals have electric and luminous organs under nervous control. In the case of voluntary action the delay in the central nervous system may be very long, but there is always some external motive for a voluntary action. The nature of reflex and voluntary action will be made clearer by a few examples. All but the last are reflexes. The first three are entirely independent of will or consciousness. They are performed by smooth muscle.

	Action	*Receptor*	*Effector*
1.	Speeding up of the heart on increasing its supply of blood.	Nerve-endings in right auricle.	Heart muscle.
2.	Contraction of pupil in strong light.	Retina of eye.	Smooth muscle of iris.
3.	Reddening of skin after a scratch.	Pain spots of skin.	Smooth muscle of small vessels which open.
4.	Secretion of saliva on smelling food.	Olfactory organ in nose.	Salivary glands.
5.	Knee jerk.	Nerve-endings in tendon.	Extensor muscles of thigh.
6.	Blinking on eye being struck at.	Retina of eye.	Eyelid muscles.
7.	Breathing.	Respiratory centres in brain.	Muscles of chest and diaphragm.
8.	Sneezing.	Nerve-endings in nose.	Muscles of chest and diaphragm.
9.	Answering a bell.	Organ of Corti in ear.	Leg muscles.

The fourth involves consciousness but not will. It can, however, be influenced by voluntary attention. The next four are performed by striped muscle, and are partly under voluntary control. The last is a very simple voluntary action. The line between reflex and voluntary action is not sharp. Only an experienced schoolmaster can tell voluntary from reflex coughing. It will be seen that the receptor organs are generally, but not always, sense-organs, that is to say, their stimulation produces consciousness as well as reflex action.

We can learn a great deal about the properties of nerve by taking a muscle with its motor nerve out of a recently killed animal. If we stimulate the nerve by electrical or chemical means, or mechanically (e. g. by pinching), the muscle will contract. This irritability continues for many hours, though the muscle is very easily fatigued unless it has a proper oxygen supply. The muscle may be made to work a lever which writes on a moving sheet of paper, and the effects of different stimuli can thus be compared. We can also measure the heat or electrical changes produced. By such means we learn the following facts about nervous conduction.

Each fibre conducts independently of the others. It conducts not a steady stream, but a series of nervous impulses. An impulse is not an electric current, but an activity of the nerve-fibre producing an electrical effect and a little heat as it goes along.

It travels at about 30 metres per second or 70 miles per hour in man. Thus a man, hit by a car going at 80 miles per hour, will probably feel nothing because his brain is destroyed before any nervous impulses from his skin reach it. After the passage of an impulse the fibre needs a rest of one-thousandth of a second or more before it can transmit another.

All impulses in the same fibre are normally of the same intensity. Thus, from this point of view, we may compare a nerve to a bundle of telegraph wires, down which electrical waves of the same intensity pass at varying intervals, but not to a bundle of telephone wires, in which the intensity of the waves is variable. Finally, the energy of a nervous impulse is so small that about four million impulses (which would take several hours to pass) would be needed to heat a nerve 1°C.

A voluntary muscle responds to one nervous impulse by a twitch, to a rapid series by a steady contraction. A contracting human muscle is getting about forty-five impulses per second. The energy liberated in a gram of contracting muscle is several hundred thousand times greater than that in a gram of the nerve which supplies it.

This ratio is about the same as that of the energy developed by a thirty H.P. motor-car running for twelve hours, to that used by the man who turns the starting-handle for a minute. Muscles can be made to contract by weak artificial electric currents as easily as by those produced by the nerve. We are not yet sure of the details of how a muscle contracts, though it seems that the lactic acid formed causes microscopic fibrils to contract, as many proteins do when placed in weak acid.

A muscle is not a heat engine, for it has a very high

efficiency such as is only found in heat engines one part of which is very much hotter than the other. Its chemical energy is converted directly into work without first passing into heat. The actual process of contraction may have an efficiency of 90 to 100 per cent., but an amount of energy greater than the work done in contraction is wasted as heat during the re-synthesis of the lactic and phosphoric acids, so that the whole process has an efficiency of only about 40 per cent.

Moreover, a good deal of energy is needed for the extra breathing and heart action during exercise, besides the basal metabolism which goes on all the time. So, considered as a machine, a man never has an efficiency of more than 25 per cent. At best, he turns three times as much energy into heat as into work. Moreover, a muscle heats up while keeping up a steady contraction, as in standing, supporting a weight, or pushing at a closed door.

Striped muscles become quite flabby when their nerves are cut, but heart muscle and smooth muscle remain active. Striped muscles have probably only one set of motor nerves which make them contract. Involuntary muscles have two sets: stimulation of one set causes increased activity, of the other set rest or lessened activity.

The effects of stimulating a nerve-fibre depend mainly on its connexions in the body, to some extent on the quantity and rhythm of the stimulus, but not at all on where in its course it is stimulated. These facts were first discovered by Müller in 1826. Thus a blow on the 'funny-bone' or just above it is felt in the ring and little fingers because the nerve from them runs near the surface at this point, and irritation of the nerves in the stump of an amputated leg will give a man pain which he feels in toes that he may have lost forty years ago.

Again, when the nerve to the face muscles has been destroyed, the power to move them may sometimes be regained by grafting the nerve supplying certain neck and shoulder muscles into the old track of the facial. But when connexion has been made the patient, in order to move the face, must will to move the shoulder.

The most accessible receptor organs are those of the skin.

They can easily be studied in an area where they are scattered, as on the side of the knee. If the skin is shaved we find that only parts round the hair roots are sensitive to gentle touch with a bristle. Each root is surrounded by a network of nerve-fibres which are easily stimulated. The small hairs act as levers, and render the 'touch spots' more sensitive. In hairless parts, such as the palm, sole, and lips, there are special receptors for touch.

Similarly if we go over the skin with a warm blunt metal point, we find that warmth is only felt at a second set of points (not those sensitive to touch), cold at another set, and pain at a fourth. Many areas on the thigh are quite insensitive to pain, as they have no pain-spots. It is characteristic of receptor organs to be specially sensitive to one kind of stimulus, which may be physical, as with the skin organs, or chemical as with those of taste and smell. They will, however, generally respond to inappropriate stimuli, if these are strong enough. Thus mustard will stimulate first the heat and then the pain-spots. A blow or pressure on the eye will make one see stars, and so on. On the other hand each receptor organ, with the paths leading from it to the brain, can generally only give rise to one kind of sensation. This, however, depends on the part of the brain to which it leads, not on the organ which is stimulated. There is no fundamental difference in the nature of the impulses in different nerves, as there is in their effects. If we stimulate the optic nerve, even after the loss of the eye, we get visual sensations, and so on.

If we put the right hand into hot water, and the left into cold for a minute, and then both into lukewarm water, this feels cold to the right hand and hot to the left. This is characteristic of the senses. They tell us more about differences of intensity in their stimuli than about their absolute intensity. When we look at a candle in sunlight we find it hard to believe that we can see by it, or even be dazzled by it, at night. We go into a dark room, and see nothing at first, but soon adapt ourselves, and see the things in it instead of a uniform blackness. After half an hour in a sound-proof room one finds the noise of one's own heart and breathing unpleasantly loud, though normally one cannot hear them. The fineness of discrimination for touch depends mainly on the closeness of touch-spots. Thus, on the palm, where they are very numerous, we can distinguish two points from one if they are 1 centimetre apart. On the back, where there are few touch-spots, this distance must be increased seven times or more.

Under the skin are receptors of many kinds. Some respond to deep pressure, and others to pain, but when once the skin is cut through, most healthy tissues are almost insensitive to pain. They become tender, however, when inflamed. The gut and other hollow organs are insensitive to cutting or burning (stimuli to which they are not normally exposed), but very painful when stretched either by unusually bulky contents or unusually strong contractions of their muscles.

There are also receptors in muscles, tendons, and joints. These send impulses to the central nervous system, which inform it of the relative positions and movements of different parts of our body. Receptors and nerves with this function are called proprioceptive, whilst those whose stimuli come from outside are called exteroceptive. Proprioceptive organs may affect the consciousness.

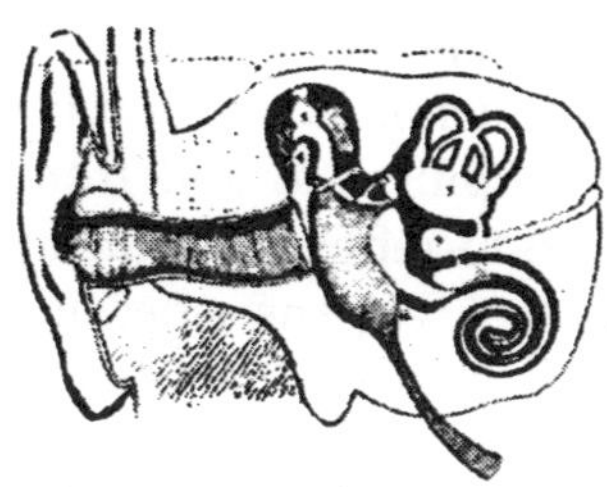

Fig. The Course of Nerve-impulses Concerned In A Spinal Reflex.

Thus we can tell how much our knee is bent even with our eyes shut, owing to the jointorgans, or how great a weight we are holding, owing to the muscle-organs. But far more important is the aid they give us without our knowing it, in the co-ordination of muscular movement. If proprioceptive impulses cannot reach the brain from the legs, as happens in a disease of the spinal cord called locomotor ataxy, in which sensation is not lost, the patient cannot co-ordinate the movements or postures of his leg muscles. In walking he raises his foot too high and brings it down too hard. He cannot stand with his eyes shut. There is no weakness of the muscles, but they cannot be used properly, as the brain gets no information as to what they are doing except through the eyes.

The spinal cord consists of a central core of nerve-cells, called the grey matter, surrounded by millions of fibres mostly running lengthways, and called the white matter. Both include a scaffolding of supporting cells. Between each pair of vertebrae a nerve leaves the spinal canal on each side.

It enters the spinal cord by a dorsal and a ventral root. The ventral root fibres go to muscles and glands, and are only traversed by impulses going outwards. The dorsal root consists of fibres carrying impulses from receptor organs to the cord. So if a dorsal root is cut we lose the capacity for feeling with a certain area of the skin, while injury to ventral roots leads to paralysis of muscles.

But most nerves contain both sensory and motor fibres, so when a nerve is cut both movement and sensation are lost in the area which it supplies. The separation of the roots serves to bring all the sensory fibres to one cell-area within the cord, all the motor fibres to another.

A nerve's only function is to conduct, but the spinal cord not only conducts impulses to and from the brain with its fibres, but gives rise to reflexes by means of its nerve-cells. This is shown by what happens when it is divided. If a man breaks his spinal cord in the neck, he dies because his breathing muscles are cut off from the brain, and get no nervous impulses to make them work. If it is broken lower down he may live for some time. He has absolutely no feeling in the parts of his body and no voluntary control over the muscles whose nerve-supply comes from the part of the cord below the break.

But if we examine him six months after the accident we find reflexes occurring in the lower part of his body. If, for instance, we pinch his foot it is drawn upwards without his knowledge or will. If the lower part of the cord is destroyed or the nerves to it cut, all reflexes cease. A great deal has been learnt about nervous activity from the study of spinal reflexes.

They are easily studied on the carcass of a frog whose brain has been destroyed by poking a blunt wire into it from behind. If we irritate its skin its hind leg scratches near the place irritated, but its responses are clumsy, and it does nothing without some fairly violent stimulus.

Provided they are in nervous connexion with the brain or spinal cord, the muscles of a limb are never quite flabby. They are mostly in a state of gentle but steady contraction or 'tone', so as to keep the limb in a definite posture. If then the knee is to be bent, as the flexor (bending) muscles of the thigh contract, its extensor muscles must relax. If they relaxed too little or too slowly there would be a strain and a waste of energy. If they relaxed too quickly or completely the movement would proceed too far, and the kneejoint might be dislocated.

Exactly the same applies elsewhere. Almost every muscle in the body, those of the trunk, jaws, and eyes, as well as the limbs, has an antagonist, and arrangements must be made for one to relax as the other contracts. As there are no inhibitory nerves to striped muscles, this can only be done by inhibiting or switching off the activity of those cells in the central nervous system which are sending impulses to the muscle which has to relax.

An idea of some of the connexions concerned in a simple spinal reflex. An impulse enters the cord through a fibre in a dorsal root from a pain-spot in the foot. The fibre divides and its branches end near nerve-cells in the grey matter. One of these cells is represented sending a fibre to a flexor muscle of the knee, another sends a fibre to an extensor. When the pain-spot is stimulated the impulses passing along it cause more nervous impulses to be generated in the cell connected with the flexor muscle, less in that connected with the extensor, so the knee tends to bend, and the foot to be withdrawn. Actually things are far more complicated.

Stimulation of a single pain-spot will only cause movement after a long time or never, and, if movement occurs, hundreds of nerve-fibres will be conducting impulses at once. To get a prompt movement one must stimulate a number of spots or fibres from them at once, as when one treads on a hot coal. Though the type of connexion have actually been observed with the microscope, in most reflexes the excitation has to pass through several neurons before it reaches the cell whose axon is the nerve-fibre to the muscle or other effector.

We must now study the function of the spinal cord in conducting nervous impulses in both directions between the body and brain. In a mixed nerve like the sciatic in the thigh, all sorts of fibres with different functions run together. Some are carrying impulses to the muscles, others from the skin and deep receptors. As they enter the cord they are sorted out according to which way they conduct, and later on a further sorting process occurs. For example, the impulses which on reaching the brain give rise to sensations of temperature, run up the spinal cord by a different path from those which give rise to sensations of touch.

These paths have been located by several different methods, which have also been applied to the study of the brain itself. First, symptoms are observed in patients, and after their death local injuries of the spinal cord due to splinters of metal or bone, burst blood-vessels, or tumours, are found. When the same symptoms are observed in another patient they can often be relieved by the surgeon owing to the knowledge so gained.

To refuse leave to examine the body in such a case is to condemn some one else to die with those symptoms. Again, after destruction or division of some parts of the nervous system one can observe with the microscope the death and degeneration of groups of nerve-fibres. Now we know that when a fibre is divided, only that part dies which is separated from the cell-body and nucleus of the neuron to which it belongs. So we can discover in which direction the cell-bodies of any bundle of fibres lie.

But in the central nervous system the long fibres always conduct nervous impulses away from the nucleus, so we discover the direction in which nervous impulses run in the fibres we have cut. Finally, we can try the effect on an animal of cutting or stimulating some part of its nervous system. (The lower parts of the system will still work after the animal has been made unconscious by an anaesthetic or by removing its cerebrum.) By such methods we can distinguish five main pairs of ascending fibre-tracts in the human spinal cord, besides numerous smaller groups.

Two of these go to the cerebellum, and their injury does not affect consciousness, but causes unsatisfactory movements and postures, the brain being without information as to what the muscles are doing. Two of them send impulses only to parts of the brain concerned in consciousness. One serves both purposes.

Before we study the functions of the brain it will be convenient to deal with the special sense-organs in the head which communicate with it directly and not through the cord. The organs of the chemical senses, taste and smell, are found in the mouth and nose. They work together, and much of the sensation we commonly regard as taste includes an element of smell.

With the eyes and nose tightly shut, taste will not distinguish an onion from an apple. The tasteorgans mostly lie in the papillae which roughen the upper surface of the tongue. There are four elementary kinds of taste, namely: salt, sweet, sour, and bitter. Other tastes are combinations of these. Each elementary taste has different end-organs. Thus, we taste sweet things best with the tip of the tongue, bitter with the back.

The end-organs of smell are a little patch of about onequarter of a square inch of yellow epithelium at the top of the internal cavity of the nose. The corresponding area in a dog is ten or more square inches, in a large shark 24 square feet. In man, smell is an unimportant, almost vestigial sense, but in the dog and many other animals it is the most important of all. So the dog's world is mainly a world of smells. But even in man it is the most delicate of the senses.

We can smell mercaptan at a dilution of 1 milligram in 20,000,000 litres of air. As about 1 cubic centimetre at a time is in the olfactory part of the nose, this means that we are affected by one twenty-thousand-millionth of a milligram, whereas the smallest object we can see with the naked eye is about a million times as large. In ordinary breathing most of the air goes past the olfactory cavity. In sniffing, some is sucked into it. No satisfactory classification of smells has yet been made.

The external ear, which is only found in mammals, is of little use to man, though some beasts can turn it so as to collect sound. In the internal ear, which lies in the thickness of the skull-wall, are the organs of hearing and balancing.

The ear-drum lies across a passage leading from the outside to the throat, and corresponding to the first gillslit of a fish. The inner two-thirds of this passage are called the Eustachian tube, and serve to equalize the pressure on the two sides of the drum.

If it is blocked by a cold the pressure becomes unequal, the drum is too much stretched to vibrate properly, and deafness results. The internal ear lies in a long 'labyrinth'. The organ of hearing consists of a tube called the cochlea, coiled like a snail's shell; that of balance consists of three 'semicircular canals' and two smaller cavities, all communicating and filled with fluid. They are surrounded by further fluid which separates them from the skull.

The drum is set in motion, like the diaphragm of a telephone receiver, by sound-waves in the air. It transmits this motion through a chain of three small bones to a membrane covering a tiny oval window in the bony labyrinth. The bones serve to concentrate the energy from the drum on to the window, which is one-thirtieth of its area. This is necessary if air-movements are to be transmitted to a watery fluid, which, being denser, is harder to move.

When the oval window is pushed in another membranous window bulges out, and it is clear that the sound-waves in the fluid must travel between them. The only path lies through the hearing organ in the cochlea. This includes a series of about 10,000 fibres (not nervous) of varying length and probably of varying tension, stretched across the tube of the cochlea, and joined by a fine membrane. Several fibres from the auditory nerve end in receptor organs on each of them.

It seems that each will vibrate to one note only, like a wire in a piano. If we play a chord loudly near a piano, the wires corresponding to the notes start vibrating. Similarly in the cochlea each fibre responds to one note and excites the corresponding fibres in the nerve. Very pure musical notes only excite few fibres, but generally they are accompanied by overtones which give them their timbre, and excite the cochlea in several places.

In ordinary noises and the vowels of speech the mixture of tones is still more complicated. Thus every sound is translated into a series of impulses travelling to the brain along a certain number of fibres of the auditory nerve, and we judge of the quality and intensity of the sound according to which fibres are excited, and how frequently. Its direction is judged mainly by the different intensities with which the two ears are excited.

The balancing organ consists of two parts. Two of its cavities contain tiny lumps of calcium carbonate called otoliths, [1] which are supported by 'hair-cells' in which nervefibres end. According to the fibres excited at any moment by the otolith pressing on the corresponding hair-cells we judge what objects are vertical, though here we are helped by our other senses.

If we lean our head the otoliths roll on to a new set of hair-cells, a new group of fibres is excited, and we alter our opinion as to what line in our head is vertical. The reflexes excited by these organs are more important than the sensations. If an acceleration of our body, as on a swing or merry-go-round, moves the otoliths from the bottom of their cavities, we get a false idea of what is vertical, but we perform the right reflexes, and lean so as not to fall. The otoliths in fact behave like plumb-lines in our heads.

Even such simple animals as jelly-fish have otolith organs to enable them to swim the right way up, and they are generally found in animals which have to balance. Some shrimps put particles of sand into these organs with their claws when they moult. If given iron filings they put these in instead. When a magnet is now held over them, the filings press upwards, not downwards, and the shrimps swim upside down. On each side one semicircular canal is horizontal, and the other two are in vertical planes at right angles. If we spin round, the fluid in one or more of them is left behind (like the water in a glass which we spin suddenly) and therefore moves relatively to the head.

Fig. Diagram of the Human Ear

In doing so it presses on microscopic 'hairs' projecting from cells, and excites nervefibres running to the brain. After we have spun for some time the fluid moves with the head (as does the water in the glass), and goes on moving after the head has ceased to spin, giving the illusion that we are spinning the other way.

In ordinary giddiness in a horizontal plane there are rapid reflex movements of the eyes which can easily be seen in others, and make things appear to spin round us. These reflexes are of great value in ordinary life as they keep the direction of our gaze fixed when we turn our heads quickly. We can become giddy in a vertical plane by turning round with the forehead or ear resting on a stick, and then raising the head. The violent reflexes of the limb and trunk muscles, which normally keep us from falling, now make us fall.

In fish the hearing and balancing organs are a specialized part of a system of canals under the skin and opening by occasional pores, which enable the fish to appreciate movements of the water round it, and its own movements relative to the water.

The eye is enclosed in a tough capsule, transparent in front only, and pierced behind by openings for nerves and blood-vessels. It can be turned by six muscles which run between its capsule and the skull. A section from front to back passes through the following structures:

- The window or cornea;

- A chamber containing a watery fluid;
- The iris, a ring of muscle, with pigment to keep out light, which regulates the amount of light reaching the back of the eye;
- The lens, a horny body whose shape can be slightly altered;
- Chamber filling most of the eye and containing a transparent jelly;
- Fine membrane, the retina, which consists of several layers of nerve-cells, and is sensitive to light;
- Layer of cells containing dark pigment, which acts like the black lining of a camera, and prevents light which has once entered the retina from being reflected;
- The tough white coat which envelops most of the eye, and helps, with the aid of the internal fluid pressure, to keep it to a definite shape.

The general structure is like that of a camera with its lens, diaphragm, and sensitive film; and the cornea and lens have such refractive indexes and curvatures that the images of external objects can be accurately focused on the retina.

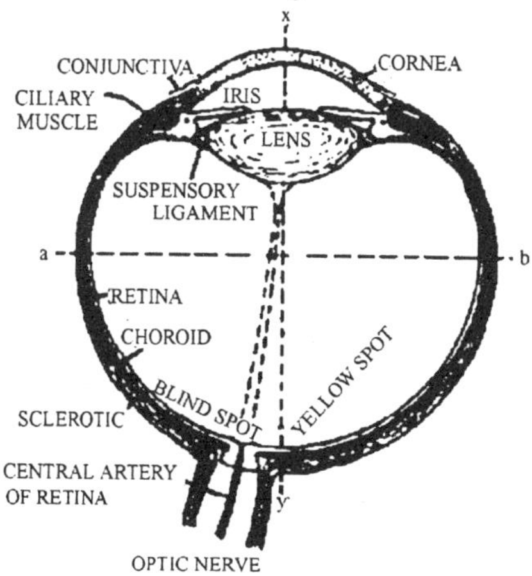

Fig. Diagram of a Horizontal Section Through the Right Eye of a Man

The lens and iris separate the anterior chamber (filled with aqueous humour) from the posterior chamber (filled with the jelly-like vitreous humour). The shape of the lens can be altered by the ciliary muscle pulling on the suspensory ligament. The image is upside-down and right-side left. If we shut the right eye, and press the outside of the left eyeball through the eyelid, we see a dark spot with a bright border well out to the right or against the nose, which appears to move up as we move the pressing-point downwards. We are stimulating the part of the retina used for looking out to the right.

The focus of the eye is altered in birds and cold-blooded vertebrates by moving the lens bodily backwards and forwards as in a camera. In man and other mammals, however, the lens is fixed, but its shape can be altered. When we wish to look at an object near to us a circular muscle round the lens contracts, and it becomes more nearly spherical.

The rays of light are therefore more bent in passing through it, and brought to a focus on the retina. If this muscle is not contracted they come to a focus behind the retina, and we see indistinctly. When we look at distant objects the directions of gaze of the two eyes are parallel, but when we look at a near one they have to be converged by the muscles which move the eyeball from outside, or we see double. The impulses coming to the brain from the eye-muscles help us to judge distance accurately. This is why it is very hard to hit a near object accurately from the side with one eye shut, as any one can easily prove for himself.

If the focusing of the eye goes wrong it can often be corrected by spectacles. When the eye is too long the rays from distant objects converge in front of the retina, and we are near-sighted. This is corrected by using concave lenses. If the eye is too short the rays from near objects converge behind the retina, and long sight results, necessitating the use of convex lenses for reading and fine work. If the cornea is more curved in one direction than another, like the bowl of a spoon, we cannot focus two perpendicular intersecting lines at the same time. This condition, which is called astigmatism, can be remedied by using lenses one side of which is a segment of a cylinder.

The iris contracts if strong light is flashed on to the eye, and expands in the dark, thus shielding the retina from too sudden changes. The microscopic receptor organs in the retina are called the rods and cones, from their shape. The rods are used for seeing in the dark, and do not distinguish between colours; the cones for vision in daylight. The diameter of a cone is about 2·5 ì, or 1/400 of a millimetre, so that there are over fifty million in each retina. We cannot distinguish two objects if their images fall on the same cone, as happens if the angle subtended by them at the eye is much less than a minute (the angle subtended by a halfpenny at 100 yards), however good our focusing may be.

There is one spot on each retina, called the yellow spot, on which we focus the object at which we are looking. Vision is most accurate here in light (but not in darkness), and becomes dimmer as we pass away from it, until with the edge of our retina we cannot tell the form or colour of things, though we can see if they are moving.

On one side of the yellow spot the nerves and the blood-vessels enter the eyeball and there are no rods or cones, so vision is absent here, since the optic nerve is no more sensitive to light than any other nerve. The existence of this blind spot can be demonstrated by putting two small objects such as halfpence on the table at a distance of about six inches and equidistant from the body. On shutting the left eye and looking fixedly at the left-hand object, meanwhile gradually approaching the head from a distance of a yard or so, the right-hand one will disappear when about two feet away.

This fact interested Charles the Second, who used to amuse himself with it until he grew so expert that he could 'take off' the heads of his courtiers. The retina contains two layers of nerve-cells besides the rods and cones, and many fibres, the last relay of which runs into the brain as the optic nerve. Compared with the ear, the eye is much better at judging the direction of the waves which stimulate it, but is sensitive to a much smaller range of wave-lengths.

The longest wave-length that we can see in the red is about 0·0008 mm. (0·81), the shortest in the violet about 0·0004, so we can only perceive a single octave of the possible vibrations, the shorter invisible ones being ultra-violet and X-rays, the longer heat and 'wireless'. On the other hand we can perceive sound-waves from 20 metres down to about a centimetre in length, a range of eleven octaves, seven of which are used in music. Moreover, the ear is better than the eye at analysing mixed vibrations. It is easy to analyse a chord into two or three notes, but we cannot tell without a spectroscope whether a given yellow is pure or due to a mixture of red and green light.

Many animals have eyes working on quite a different principle from ours, namely, built of little units each looking out in one direction , just as each of our 'cones' receives light from one direction only. Hearing organs are found in a few insects, often in their legs or bellies, and are provided with drums and hair-cells similar to our own.

We must now consider the brain. The human brain is built on the same plan as the frog's, but one part of it (the cerebrum) has grown in man and related animals to be much larger than the rest. It is on this part that the main differences between the behaviour of a man and a frog depend. We shall first consider the lower parts of the brain, which are not so very different in the two species. As the spinal cord enters the head it expands into the medulla oblongata, in which there are nerve-centres governing the involuntary activities of the body.

For example, if an animal's head is cut off and the blood-vessels of the neck tied to prevent the loss of blood, it will not breathe, nor if we obstruct its aorta will the heart slow down. But if only that part of the brain above the medulla oblongata is destroyed, both these processes continue, though the breathing is clumsy. The medulla also regulates such functions as digestive secretion, including salivation; movements of the digestive organs, such as peristalsis and vomiting; and a number of reflexes in the circulatory system to be described later. The involuntary muscles and glands concerned are controlled through the autonomic or involuntary nervous system. This consists of two parts, the sympathetic and the parasympathetic. The latter system is composed of the vagus nerve which runs down the neck from the medulla, and supplies the chief involuntary organs from the thyroid gland down to the beginning of the large intestine; a few small nerves to glands and involuntary muscles in the head; and some nerves leaving the lower end of the spinal cord for such organs as the large intestine and urinary bladder.

There are two fundamental differences between involuntary and voluntary motor-nerves. The latter run straight to their destination, the former end in a ganglion where each excites one or more nerve-cells from which fibres run on to the muscle or gland. Also a single nervous impulse down an involuntary nerve has no effect. Several are needed to excite the rather sluggish organs which they supply. Most viscera get fibres both from the sympathetic chain and the parasympathetic system, and the two systems are generally antagonistic.

Thus stimulation of the vagus slows down the heart, while the sympathetic speeds it up. The vagus makes the stomach and gut move and secrete their juices, while their sphincters such as the pylorus relax; the sympathetic diminishes their movement, secretion, and blood-supply, and tightens their sphincters. In other words the one promotes, the other hinders digestive activities.

During violent exercise impulses pass down from the brain to the heart and gut; these set the heart beating faster and stronger, drive blood out of the vessels of the gut, and slow down the gut's movements. The autonomic nerve-trunks also contain afferent fibres, but not very many, as the brain does not need very detailed information about events in the viscera.

Most of the nerve-fibre groups in the cord pass through the medulla, though some of the ascending ones end round neurones there whose axons pass on to higher parts of the brain; so that the medulla acts as a relaying station for the impulses which they carry. In the medulla, too, most of the fibre-tracts to and from the higher part of the brain cross to the opposite side of the body. Hence the left side of the brain is concerned with nervous impulses to and from the right side of the body. The paths to and from the cerebellum, however, are mainly uncrossed.

Above the medulla lies the mid-brain. This contains the

nerve-cells whose axons form the motor-nerves to the eyemuscles, but its most important functions seem to be in connexion with posture. If in an animal all the brain above the mid-brain is destroyed, it goes into a rigid state with the legs thrust out and the trunk stiff as in standing.

Just as the spinal cord alone or along with the medulla will organize reflex muscular movements, so with the mid-brain in addition reflex posture is possible. Thus a 'decerebrate' animal, i.e. one in which the cerebral hemispheres have been removed, though unconscious, can to some extent adjust its standing posture.

If its head is bent down, it bends its forelegs, and so on. Similarly, if in a man the nervous pathways from the cerebrum are destroyed, certain groups of muscles cannot be moved voluntarily, but remain contracted in a state called spastic paralysis so long as the brain-stem is acting on them, whereas if the injury to the nerve-paths is lower down the same muscles are equally paralysed, but flabby.

Behind this part of the brain is the cerebellum, an organ with several layers of nerve-cells on its outside, and a few large groups of cells inside, all connected up by numerous nerve-fibres which run between them and to other parts of the brain.

When it is damaged there is no loss of sensation or power of thinking, but there is a loss of muscular tone, and a great deal of jerkiness and inco-ordination of movement. It bears the same relation to the proprioceptive system as the cerebrum to the exteroceptive. All the impulses from muscles, tendons, joints, and labyrinth are co-ordinated there so that in a movement or posture the right muscles may be contracted to the right extent at the right moment.

The rest of the brain without it is like a general who gets inadequate reports of the movements of his own troops. If a man with cerebellar disease tries to grasp an object, he moves his hand in a series of jerks and grasps in the wrong place.

Above these organs is the cerebrum, which, in man, but not in other animals, is many times larger than the rest of the brain. The human cerebrum contains more than a thousand million nerve-cells each connected by fibres with scores or hundreds of others, so we can get some idea of its complexity by imagining a telephone exchange in which the whole human race were acting as operators. The main mass of nerve-cells lies on the outside, and this 'grey matter' is folded to increase its area. In the middle, at the top of the brain-stem, are two masses of nerve-cells called the optic thalami, into which run the optic nerves and also fibres from below which carry up impulses from all the other sensory nerves on the opposite side of the body.

A child born without cerebral cortex but with thalamus lived for three years without showing as much signs of consciousness as a normal baby a few days old.

On the other hand a dog without cerebral cortex can walk about, though it runs into obstacles; but it shows no signs of recognizing anything, and food has to be placed in its mouth before it will eat. Judging from cases of disease in man, a dim kind of consciousness seems to be associated with the thalamus. When all the paths leading upward from it on one side are destroyed, the sensations on the opposite side of the body are abnormal.

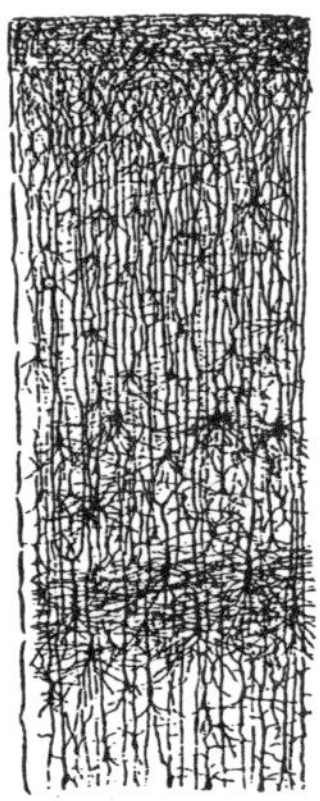

Fig. Microscopical Section of Part of the Human Cerebral Cortex

Light touch is not felt, but a slight scratch is felt as a horrible pain which cannot be localized, and the touch of a warm bottle as a huge pleasure. The more complicated senses, such as vision, are not represented in the thalamus, though it acts as a relay for visual impulses. Different parts of the cerebral cortex have very different functions, and about half of it has been properly mapped out as regards its function. Gentle stimulation of certain areas gives rise to movement, generally of muscles on the opposite side. This is always fairly well co-ordinated, i.e. a number of muscles work together, as is the case in voluntary action.

The size of the area devoted to a group of muscles depends on the complexity and delicacy of the movements required of them. Thus the tongue has a cerebral area as large as the whole trunk, and the eye-muscles an area about a third as large as all the other muscles put together.

When a motor area is irritated, as by a splinter of bone or a small tumour, a special type of epileptic fit results, in which the involuntary muscular movements begin in the muscles governed by the irritated area. A knowledge of cerebral localization renders a surgical cure of fits of this kind possible. When part of the motor area is destroyed the corresponding muscles are paralysed, though later on other areas may partially take its place, and some voluntary control return.

In front of some parts of the main motor area, especially on the left side, is a region whose injury in man causes, not paralysis, but a failure of the more complex movements, such as those involved in speaking and skilled manual operations. Finally,we must remember that when we are 'doing nothing' the cortex is all the time inhibiting the postural centres in the brain-stem from producing rigidity, so that a voluntary movement may sometimes merely be a stopping of this inhibition.

The main motor area and the region just behind it constitute the sensory area for all the senses except the special senses of sight, hearing, taste, and smell. It is gradually being mapped out, and war injuries gave us a great deal of information. The sensory area for each part of the body includes the corresponding motor area and an area behind it. The hands have a very large proportion of the whole.

If the brain is injured a little behind the sensory area, the sensations are still felt but cannot be put together. For example, a man can state just what part of his hand is being touched, and whether any finger is bent or not, but he cannot say what sort of object he is holding in his hand. This part of the brain is therefore concerned in putting sensations together and interpreting them.

The optic nerves join before reaching the brain, and half of each crosses over, so the left side of the brain gets fibres from the left side of each retina, both of which look out on the right. So if the left visual area is destroyed a man can see nothing to his right with either eye. Similarly the different parts of the field of vision are represented on the visual area. If the visual areas of the brain are destroyed a man is quite blind, but may retain a good deal of visual memory; but if the neighbouring areas are destroyed, this too is lost.

In thought and speech a great many parts of the brain are employed at once. Thus to fully understand the meaning of the word *apple* we require memories of sight, hearing, smell, taste, and touch, and the power of co-ordinating them. Some parts of the cortex are supplied entirely with fibres from other parts and clearly serve as centres for association and co-ordination.

We are gradually finding out the functions of these parts by studying the effects of wounds and the degenerative changes found in the brains of the insane. In right-handed people the left cerebral hemisphere generally contains the main speech-centres (and vice versa); injury of a large area of this will cause failures in speech, and in the thought behind it.

According to the part injured there may be a mere slurring of words with apparently fairly clear thought, an inability to remember the names of things, or a failure to construct sentences and to think out problems. But the co-ordination of the brain-cells is less understood than that of other organs.

We can only emphasize the very important part played by inhibition, that is to say, the checking by one part of the brain of the activities of another part. In the spinal cord one reflex can inhibit another. For example, in a decapitated animal a stimulus, which would be painful to an animal with a head, at once inhibits reflex scratching movements.

Voluntary attention means an inhibition of all our mental activities but one, and resistance to temptation is an inhibition of our more primitive activities, such as eating or losing our temper.

Many pathways are known for nervous impulses from the brain down the cord to the motor-cells in its grey matter whose axons form motor-nerves. One leads from the motor areas of the cerebral cortex to the opposite side of the cord, and is concerned in voluntary movements.

Others descend from the mid-brain, which is under the influence of the cerebellum, to the opposite side of the cord, and are mainly concerned with posture and muscular tone. Another leads from the medulla, and is concerned with rapid reflexes to stimulation of the labyrinth, i.e. to keeping one's balance.

Chapter 6

Organic Regulation

The nervous system serves to co-ordinate the activities of the different organs to some extent, but it is not in itself essential for the life of the tissues. A leg will live for years without nerves, but only for an hour or less without blood or some artificial substitute for blood. The cells in a higher animal are like skilled workmen, very efficient at their own job, but not at other jobs.

Thus a single cell in Hydra may serve for protection, be sensitive to external stimuli, contract when stimulated, pass on excitation to its neighbours, and perhaps secrete mucus, but in none of these ways will it act as efficiently as the various cells of a mammal, each of which performs one of these special functions.

The latter are enabled to specialize largely because they have a nearly constant environment, constant in chemical composition and temperature, and do not spend any energy in adapting themselves to change in it. This environment is supplied by the fluid part of the blood. The factors in the internal environment, and how they are kept steady or adapted to new conditions. Most of the general symptoms of disease are due to upsets of the internal environment.

Each organ must have food, oxygen, and a means of getting rid of waste products. But this is not all. It must have them in the right amounts. Too much oxygen is just as deadly as too little. And it must also have in the right amounts other substances which it does not use for work or repair. An animal dies if we halve or double the amount of potassium salts in its plasma, though it does not turn potassium salts into anything else. So these too have to be kept steady. Further, an organ may need different amounts of a given substance at different times. If a muscle suddenly starts work, its O_2 -consumption and CO_2 -production increase about fifty times. It was already using most of the oxygen in the blood which passed through it, so it must increase its blood-supply correspondingly. When it is at rest about five out of every six of its capillaries are shut.

When it begins to contract these at once open and the others open wider. The small arteries also open wider. In a resting muscle they are kept almost shut by the slight alkalinity of the blood, and also probably by the presence of oxygen. Now if we cut the leg off a recently killed frog and run fluid through the blood-vessels, they will open up if this fluid is not sufficiently alkaline, or is short of oxygen. Just the same thing happens when a muscle contracts or a gland begins to secrete. The O_2 flows from the blood and CO_2 is poured into it, making it acid; the blood-vessels relax and widen, and the organ obtains an adequate flow of blood. Other products of activity besides CO_2 probably co-operate in producing this effect.

But when any large organ opens up its vessels the arterial pressure would fall unless the heart pumped harder. The other organs would then go short of blood. We must therefore study the working of the heart. Like other involuntary muscles, it will work without any nervous control. If we take the heart out of a recently dead animal or man, and supply it with warm blood or an appropriate salt solution containing oxygen, it begins to beat again and may go on for many hours. Even an isolated piece of it will beat. In a mammal the beat starts at the entrance of the great veins to the right auricle in a special piece of tissue known as the 'pacemaker', which does not contract but stimulates the neighbouring muscle. If we warm the pacemaker, the whole heart beats faster; if we destroy it, the heart first stops, then begins to beat at a slower pace of its own.

The auricles contract almost instantaneously when stimulated by the pacemaker, but they are only connected with the ventricles by a narrow bridge of conducting tissue in which the wave of excitation is delayed for about one-tenth of a second, and then passed on almost simultaneously to all parts of the ventricles. If the bridge is damaged, as in some forms of heart-disease, the ventricle may only respond to every second or third beat of the auricles. If it is destroyed they beat at their own rather slow rate, and cannot be speeded up.

Now if we take an isolated heart, or a heart whose nerves have been cut, and give it an increased supply of blood, it can increase its output per beat but will not increase its rate at all. The rate is governed by two pairs of nerves. Of these, the vagi are one; if they are stimulated the heart slows down; if they are cut it speeds up, showing that they are normally acting as a brake on it, mainly through the pacemaker. The other pair, called the accelerators, come through sympathetic ganglia from the spinal cord. Stimulation of them speeds up the heart. Both pairs are governed by the same centres in the medulla oblongata.

If the blood-supply to the heart is increased, the great veins and auricles are distended, and receptor organs in their walls send impulses up to the brain which result in the vagus brake being slackened by a reflex action; and if the stimulus is sufficient, the accelerators are set to work. Hence, when more blood reaches the heart from the open vessels of an active muscle, it increases its rate and force.

Another set of reflexes keep the arterial pressure steady. A pair of nerves called the depressors run from receptor organs in the aorta to the medulla oblongata, which they enter with the vagus. If the aorta be distended by an abnormally high blood-pressure, impulses run up them to the medulla. The reflex response to this is a slowing of the heart by the vagus and an opening of small arteries. The opposite occurs if the aortic pressure falls. There is also a pressure-gauge in the brain itself. If pressure is put on the brain from outside, for example, by a clot of blood under the skull, its vessels will collapse, and the brain will force the heart to raise the arterial pressure till they open up again. In this way the arterial blood-pressure is kept steady, so that any organ can obtain the blood-supply it needs by opening up its bloodvessels.

But the brain does not allow an indiscriminate competition between different organs for blood-supply. The arteries as well as the heart are under nervous control. A series of nerves called vasoconstrictors run in the sympathetic system to the smooth muscle of the arterial walls. The vasoconstrictors come into play in circumstances which affect the body as a whole, such as change of posture, or violent exercise.

When a man gets up after lying down the blood tends to flow into his belly and legs, and this is prevented by the contraction of the arteries of these parts, under impulses from the vasomotor centre, which lies near the heartregulating centre in the brain. If he has been in bed some days the vasomotor centre is out of practice and his brain runs short of blood, so that he becomes *dizzy* and may faint.

If there is a great deal of blood in the guts and skin, as when we are sitting before the fire after a heavy meal, this may even happen on getting out of a chair. Again, during muscular exertion the arteries to the guts contract, and digestion has to stop. The vasoconstrictor nerves are also excited by chemical stimuli to the brain.

If we are throttled or breathe very impure air, the CO_2 of the blood goes up and the O_2 down. The vasomotor centre then narrows down all arteries except those to the heart, lungs, and brain, and the blood-pressure rises. These three essential organs must have an adequate supply of oxygen, whatever else goes short. The other organs, if left to themselves, would open up their vessels, but in a general emergency they are not allowed to do so. The heart is also speeded up.

There are also vasodilator nerves. For example, when a dog gets hot the vessels in its tongue are opened up by a special nerve. Moreover, many of the nerve-fibres whose stimulation causes pain send branches to the local blood-vessels. When a pain-spot is stimulated, impulses run up to the spinal cord. They also run directly to the local vessels, which open up, causing reddening of the skin; this is almost the only reflex in higher animals of which the nervous path is entirely outside the central nervous system.

The above are cases where an organ gets blood which is

not to be used mainly as a gas-carrier. The same occurs in an actively secreting gland. A salivary gland when active uses three times as much oxygen as when at rest. But it also needs a great deal of water to make saliva, and it is as a source of water rather than of oxygen that it needs blood.

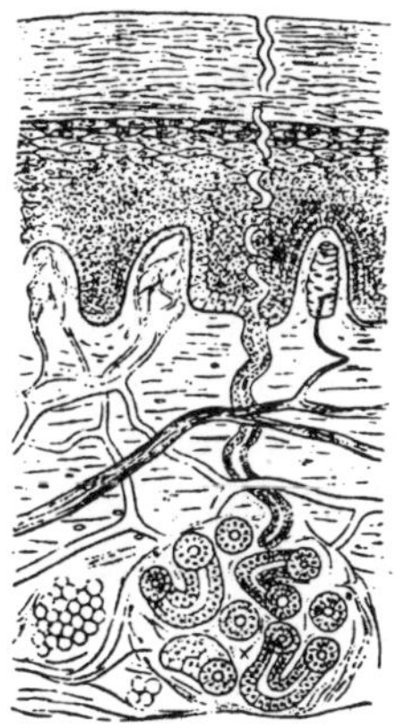

Fig. Diagram of a Microscopical Section Through the Human Skin

Its bloodsupply must go up five times, and as the ordinary chemical call for blood is not effective, a vasodilator nerve is used. As the blood-supply goes up more than the oxygen-consumption, the venous blood of the active gland is actually redder than usual.

A much more important case of non-respiratory bloodsupply is the skin's. The human skin excretes energy just as the kidneys excrete matter. All the heat produced in the body has to get out, and seldom does more than a fifth of it get out in the breath. The remainder goes through the skin, and its loss is regulated in such a way as to keep the temperature of the body very constant.

If we go into a hot room the same amount of heat has to be lost in a given time, but it is obviously harder to get rid of it. If we work hard and produce more heat in our bodies, more heat has to be lost in a given time, though the loss of a given amount is no harder. In each case the skin responds in the same way. Its small arteries open up and it gets red and warm.

Heat is thus brought from the inside to the surface in large amounts and rapidly lost, as from the radiator of a motor vehicle. If this means of losing heat is insufficient, we begin to sweat. The sweat comes from microscopic glands under nervous control. It consists of water containing less salt than the plasma. When this evaporates the skin is greatly cooled, for water has a big latent heat of evaporation. Sweat that does not evaporate does not cool us, and it cannot evaporate if the air is already saturated with water-vapour.

So sweating is useless in a hot and damp atmosphere, which is therefore far more oppressive than a dry one of the same temperature. The ordinary thermometer does not tell us whether we shall be able to lose heat or not. For this purpose we use a 'wetbulb' thermometer. The bulb is wrapped in a wet cloth, so that the drier the air is, the more heat it can lose by evaporation. It is therefore in the same position as a man whose clothes are soaked with sweat. If the air is saturated with water the wet- and dry-bulb thermometers have the same reading, if the air is dry the wet-bulb thermometer may read more than 100°F lower.

Men can stand dry heat far above boiling-point, staying in a room where a steak is cooked in five minutes, and only coming out when their hair begins to singe. But a wet-bulb temperature above 90°F is fatal, and above one of 75°F the capacity for work is lowered. We get an idea of the efficiency of sweating by considering a man in fairly dry air at body-temperature (98·5°F).

He can lose no heat by conduction or convection, so it must all be used in evaporating water. He has to lose 3,000 kilocalories per day. But the evaporation of 1 litre of water at body-temperature requires 570 kilocalories, so in a day he must sweat 5·3 litres, or 9·3 pints. Actually many men can sweat 1 litre per hour, and the world's sweating record is held by an English coalminer who lost 18 pounds (1·8 gallons, or 8 litres) in 51/2 hours.

To make up for the loss of sweat one must drink more water and eat more sodium chloride than usual. Miners working in great heat are therefore fonder than the average man of bacon, kippers, and salt. Many animals, such as dogs, have very few sweat-glands, but produce a very thin saliva which they evaporate by rapid shallow breathing, panting with open mouth and tongue hanging out.

Several parts of the brain are concerned in heat regulation. They receive nervous impulses from the skin, and also from local organs in the brain which measure the temperature of the blood like thermometers. Thus, if certain parts of an animal's brain, or the blood going to them, are heated, the animal begins to flush and sweat, and the rest of its body is cooled down. If the brain is cooled the animal shivers and its temperature rises.

During adaptation to heat we cannot cut down our heatproduction except by keeping still, but when cold, besides shutting down the skin-vessels, we first tighten up our muscles, then shiver, and finally take exercise. In all these ways more heat is produced. In many diseases the temperature rises. This is not due to increased heat-production, but to diminished heat-loss owing to perverted function of the temperature centres.

A man whose temperature is thus rising feels very cold until it has reached the new level to which he is regulating. He shivers and complains of the draught, to which he may put down his illness. If his temperature falls quickly he sweats profusely and feels very hot.

Mammals and birds have a nearly constant temperature, but other animals have a variable temperature, a fraction of a degree above their surroundings. If we warm a 'cold-blooded' animal through about 5°C we double its rate of oxygen-consumption, and all its other activities. For example, it is possible to read the temperature within 1°F by measuring the distance walked by an ant in a minute! Cold-blooded animals cannot move quickly in winter, and mostly die or rest in holes. The activities of mammals and birds are not slowed down, so they are the dominant animals in temperate and cold climates.

But in hot countries, snakes, crocodiles, and so on, are able to compete successfully with warm-blooded animals. A few mammals, such as hedgehogs and dormice, compromise by sleeping through the winter at a low temperature (and therefore a low rate of oxidation), but they never let their temperature fall to that of their surroundings.

We must now turn to chemical regulation of the composition of the blood and tissues. It will be convenient to begin with the gases, the quantity of which in the blood is regulated by breathing.

The obvious duties of the lungs are to get rid of CO_2 and let in O_2, and if we go into a room containing say 6 per cent. of CO_2 instead of the normal 0·03 per cent., or 10 per cent. of O_2 instead of the normal 20·9 per cent., the breathing increases greatly. However, a small drop in the O_2 of the air breathed has no visible effect on the breathing, because the haemoglobin is already almost saturated with oxygen at a pressure less than that in the lungs.

So want of O_2 cannot be what normally keeps the breathing going. To find out how the breathing is regulated we must take the samples of air from the very bottom of the lungs, where it is in equilibrium with the blood. This, which is called the alveolar air, can be obtained at the end of a deep breath out.

The amount of carbon dioxide in it is very constant, about 51/2 per cent., whereas the amount of oxygen varies a good deal. If the amount of carbon dioxide increases by only 3 per cent. of its normal value, the breathing is doubled; if it falls by the same amount, as after voluntary over-breathing, the breathing stops.

The main reason why we breathe more during moderate muscular exercise is because more carbon dioxide is being produced, and this stimulates the respiratory centres in the brain to make the breathing muscles do more work. Thus the lungs have the function of keeping the CO_2 pressure in the tissues at the normal level, not merely of excreting it.

Carbonic acid seems merely to act on the respiratory centre in virtue of its being an acid. If another acid, such as hydrochloric, is injected or drunk, the breathing is greatly increased, while it slows down when an alkaline substance such as sodium hydrogen carbonate is taken. The most familiar case, however, is that of very violent exercise. When the muscles are working so fast that they cannot get enough oxygen for their recovery process, lactic acid accumulates in them and leaks out into the blood, from which it is only gradually removed.

So after running a quarter-mile the extra carbon dioxide is got rid of in the few minutes of violent panting which succeed the race, but a small increase of the breathing, due to lactic acid, may persist for half an hour or so. During this time the alveolar carbon dioxide pressure is kept below normal by the extra breathing, thus compensating for the acidity which would otherwise be produced by the lactic acid.

Serious oxygen-want also excites the respiratory centre. If one goes into air containing only about half the normal 20·9 per cent. of O_2, one at once begins to pant, but after a while the panting dies down, because a lot of CO_2 has been blown out of the body by the increased breathing, and the respiratory centres have no more reason to discharge nervous impulses than before, their normal stimulus, CO_2, being reduced in quantity.

So a man who has gone into bad air at first pants enough to keep his blood supplied with oxygen. Then the breathing becomes normal, and he falls unconscious with oxygen-want. A candle is often a much better measure of oxygen-want than one's own feelings.

Another way in which the breathing is affected is by the process of digestive secretion. When the stomach secretes hydrochloric acid the blood would be too alkaline if carbonic acid were not kept back to take its place, so the breathing is slightly slowed down. Later on, the pancreas and intestine begin to remove alkali from the blood for their secretions, and to prevent it getting too acid the breathing has to be increased. These changes are too small to observe directly, but can easily be measured. We can get some idea of why the alkalinity of the tissues has to be regulated so carefully by experimenting with tissue or enzymes taken from them.

If we take a dead organ and preserve it carefully from bacteria it does not putrefy. But if it is kept at body-temperature the tissues gradually soften and are found to be digesting themselves. This is due to enzymes in them. The dying tissues produce acids, and in a slightly acid medium these enzymes work very much more rapidly than in an alkaline or neutral one.

Thus, to prevent an organ from digesting itself it must be kept slightly alkaline. The best known of these enzymes is that in the liver which breaks up its glycogen into sugar. A quite fresh liver, besides being very tough, does not taste sweet. If it is allowed to digest itself for a few days it not only becomes tender, but sweet. Some other enzymes act more rapidly in a medium more alkaline than the normal, so if the reaction of the tissues is altered the normal balance between the different chemical processes is upset, and death may occur.

Just as the lungs regulate the amount of gases in the blood, the kidneys regulate the amount of the soluble bodies. The blood which passes through these organs is always altered so as to resemble an 'ideal' blood. Thus, if there is more water in the blood-plasma than in this ideal or standard plasma the kidney secretes an unusually watery urine, and the plasma of the blood in the renal vein therefore contains less water than the arterial blood, and resembles the standard plasma more closely.

If, as is more usual, especially in hot weather, there is rather less water in the plasma than in the standard plasma, the kidney secretes a concentrated urine, so the blood leaving the kidney contains more water than that entering it. The substances found in blood and urine can be divided into two classes. The first class includes almost all foreign substances, for example, iodides, dyes, or foreign proteins injected into the blood.

These are removed by the kidneys, however little there is in the blood. It also includes some very important waste-products, such as urea, the substance which contains most of the nitrogen resulting from protein oxidation. The rate at which such substances are excreted is roughly proportional to the amount in a given volume of blood. The second class includes most of the normal constituents of plasma, such as sodium, potassium, calcium, magnesium, chloride, bicarbonate, phosphate, and sugar. These substances are only excreted if the quantity of one of them contained in a given volume of plasma exceeds a certain limit, called the 'threshold'. For example, the amount of chloride in the plasma is generally a few per cent. above the threshold, and there are, therefore, chlorides. in normal urine.

But if we drink a lot of water after violent sweating, the amount of chloride in the plasma falls below the threshold, and it disappears from the urine. Normal blood contains about 0·10 per cent. of glucose, a simple sugar. If a healthy man takes 100 grams of glucose the amount in the blood rises to about 0·13 per cent., but none appears in the urine. The threshold value for the kidney is about 0·17 per cent. If, therefore, the arrangements for storing sugar are out of order, as in diabetes, a dose of 100 grams will make the blood-sugar rise above 0·17 per cent., and sugar will appear in the urine.

Besides excreting substances found in the blood, the kidney makes a few substances. For example, sulphuric and phosphoric acids are made throughout the body by the oxidation of the proteins. The kidney has to get rid of these, but its cells and those of the urinary passages are damaged by strong acids.

It therefore excretes the sulphuric acid not as such, but as ammonium sulphate, which is neutral in reaction. But there is not enough ammonia in the blood to furnish all that is required for this purpose. Ammonia and ammonium salts are poisons when injected into the blood-stream, and the liver converts almost all the ammonia reaching it into urea, which is nearly harmless. So the kidney has to make its own ammonia, and the more acids it has to excrete the more ammonia it makes.

The kidney is doing work like a muscle, for work has to be done in concentrating substances, just as in compressing gases. For example, to concentrate the urea in a litre of blood into about 20 cubic centimetres of urine, as the kidney does every 20 minutes or so, requires at least as much work as to compress a litre of gas containing as many molecules as there are urea molecules in the blood, into 20 cubic centimetres. Actually the number of urea molecules in a litre of blood is the same as that in a litre of gas at a tenth of an atmosphere pressure, so the work needed is that required to compress this gas to a pressure of five atmospheres, i.e. 40 kilogrammetres. In order to do this work the kidney needs oxygen.

Its oxygen-consumption can be measured by determining the rate at which blood flows through it and the amount of oxygen lost by this blood. As a matter of fact, the kidney uses a good deal more oxygen per gram per minute than the heart, and like the heart, will increase its oxygen-consumption three or four times if it is given work to do.

But if we inject salt-solution of about the composition of plasma, the kidney needs no more oxygen, although the volume of urine secreted per minute is increased. This is because it has no work to do in concentrating the salt, but it merely acts like a filter. If on the other hand we inject urea or sodium sulphate, its oxygen-consumption increases, as these substances have to be concentrated. Other glands behave in a similar manner, requiring more oxygen when stimulated to do work.

Chapter 7

The Internal Environment

We saw that, among other things, the kidney was responsible for preventing the amount of various inorganic substances in the plasma from rising above fixed values, while, other organs serve to keep them up to those values. What would happen if the amounts of these bodies deviated from the normal? Their importance is shown by the extraordinary fact that an organ such as the heart or liver can be kept alive for many hours in a solution of inorganic salts which are present in nearly the same proportions as in plasma. For mammalian organs such a solution is:

$NaCl$, 0·8 per cent.; KCl, 0·02 per cent.; $CaCl_2$, 0·02 per cent.; $MgCl_2$, 0·01 per cent.; $NaHCO_3$, 0·1 per cent.

This solution, whose composition was worked out by Ringer, must be saturated with oxygen, and a little glucose may be added as a source of energy. Now if we leave out the various constituents, or increase them above the standard amounts, we shall find out what functions they are performing. The salts do not permeate the cells, or only do so very slowly, but affect their surface properties.

If the solution perfusing, say, a rabbit's excised but still beating heart is diluted with water, the heart swells up and stops. It has become sodden with water as does the skin of the hands in a hot bath, for water runs into the cells and dilutes their contents. If we make the solution too strong the cells shrivel up from loss of water.

If we replace most of the sodium chloride by the corresponding number of molecules of cane-sugar the heart goes on beating. Clearly the main function of the sodium chloride is to prevent the cells from taking up too much water. If we leave out the potassium the heart goes into a state of cramp. Potassium is needed for its relaxation.

If we leave out the calcium it stops in a flabby condition, and so on. A heart so stopped may be revived hours later by adding the missing salt. Other organs behave in the same way. Thus, if there is too much or too little calcium in the fluid in its blood-vessels, the kidney refuses to hold back sugar when the amount of sugar in the fluid perfusing it is below the normal threshold value.

We can also observe effects of the same kind on a man or animal, though here the nervous system is generally affected before the other organs. Thus, if a man drinks water for some hours more rapidly than his kidneys can get rid of it he gets cramps, and later convulsions. If he lowers the calcium to half its normal value he gets another type of cramps, and so on.

One of the most remarkable things about the plasma salts is that they are nearly the same as those of sea-water diluted with water to about three times its volume. Such a solution would, however, among other things, contain too much magnesium and sulphate.

Now the blood of marine invertebrates is very like sea-water, while that of sea-fish is generally somewhere between that of invertebrates and that of land vertebrates. Many people, therefore, think that land vertebrates are descended from fish which left the sea when it contained less salt than now, and that our blood-plasma has kept the composition of the sea-water to which the cells of our ancestors were accustomed.

Besides the inorganic substances mentioned, the plasma contains phosphates, which are also kept in by the kidney. They play a very important part in the formation of bone, which consists largely of calcium phosphate. There is an exceptionally large amount of phosphate in the plasma of growing children, and of adults who have broken a bone and or of calcium in a child's plasma falls below normal it is unable to form bone properly, and develops rickets.

Although an organ can be kept alive for many hours by inorganic substances, yet organic compounds are of course needed for its prolonged existence, and a heart will survive longer if, for example, a little sugar is added to its salt solution. As we saw, the blood contains about one part in a thousand of sugar, and this does not fall much in a starved man or animal. In this case glycogen is broken down by the liver to keep the sugar-level up, and after the glycogen is mainly used up, the organs then oxidize fat rather than sugar, and leave the blood-sugar about normal.

In the same way the amount of amino-acids remains fairly steady, though of course there is a small temporary rise after a protein meal. The amount of fat varies somewhat more, and after a very heavy fat meal the plasma may be quite milky with microscopic fat-drops.

The liver plays an important part in keeping the amounts of sugar and amino-acids steady. It stores extra sugar as glycogen, and removes the ammonia from excessive aminoacids. The residue of the amino-acid molecule left can be oxidized, and in some cases can be made into sugar if required. But besides dealing with excess of normal bloodconstituents it can destroy poisonous substances coming to it from the gut. These substances are generally the result of bacterial action there.

The bacteria attack the proteins of our food, but have not enough oxygen to utilize them fully. They therefore excrete unoxidized fragments of proteins, just as the yeast-cell excretes ordinary alcohol made from sugar which it cannot burn. Among these excretory products are phenol ('carbolic acid'), cresol, indol, and skatol. The last two are foul-smelling, and all are poisonous. They are absorbed from the gut, but on reaching the liver they are combined with sulphuric acid, apparently by an enzyme, to form quite harmless substances which are excreted by the kidney.

If, however, they enter the blood in very large amounts a proportion gets past the liver and the whole body appears to suffer. The liver acts in the same way with many other substances. It has more varied chemical functions than any other organ.

Besides foodstuffs and various colloids with which we

shall deal later, the plasma contains in exceedingly small amounts certain organic substances of great importance which are poured into it by special organs, including the ductless glands, such as the thyroid. Some of these substances, like adrenalin, are present in varying amounts and act as hormones or chemical messengers between different organs. Others seem to be present in fairly constant amounts, and are needed for the normal working of the body.

Insulin is produced by certain microscopical 'islands' of tissue in the pancreas, and passes out of them into the bloodstream. If these islands of insulin-producing tissue are removed or diseased the tissues become more or less completely unable either to oxidize or store the sugar of the blood.

The amount present in the blood increases, especially after meals containing carbohydrates, and when it rises above the threshold value the kidney excretes it and it is wasted. What is even worse, the tissues begin to make sugar from proteins, and this too is excreted. So a man with severe disease of the parts of the pancreas which make insulin gradually wastes away. This condition is called diabetes.

If he is still making some insulin we can often keep him alive on a meagre diet which he can just deal with, but in severe cases we have every day, or several times a day, to inject insulin made from the pancreas of animals. Unfortunately an overdose of insulin will depress the blood-sugar below normal, which may bring on convulsions and death. Insulin has not yet been obtained in a pure state, but we know that very little is needed, for less than one milligram per day of our strongest present preparation is needed even in severe diabetes.

The thyroid gland in the neck makes a substance called thyroxin, which has been obtained in a pure crystalline form and whose chemical structure is known. Absence or removal of the thyroid does not lead to immediate death, but to a condition called myxoedema in the adult mammal, cretinism in the young. In this state the resting O_2-consumption is only about 60 per cent. of the normal, and the adult patient becomes fat, sluggish, stupid, and bald; while a child develops abnormally and is idiotic; and a tadpole never metamorphoses into a frog. These conditions can be completely cured by administering thyroxin or extracts of the gland. Fortunately thyroxin is not destroyed, like insulin, by digestive enzymes, so it can be given by the mouth.

Thyroxin has such a powerful effect that only about a third of a milligram per day is needed to keep a thyroidless man normal, and this amount will cause the oxidation of a quarter of a million times its weight of glucose. The thyroid gland sometimes swells up and produces too much thyroxin. The resting O_2-consumption then rises and may reach double the normal amount. In one hospital the patients in the ward reserved for hyperthyroidism eat twice as much as those in any other ward! They become thin and nervous, and their eyes tend to protrude. They can generally be cured by cutting out the gland wholly or in part.

Thyroxin is an organic substance including four iodine atoms in its molecule, so if there is not enough iodine in the food and drink it cannot be made in sufficient quantity. In many inland districts swelling of the thyroid is found along with a low or normal resting metabolism. This disappears when a few milligrams a week of an iodide are given. The gland which has been over-working in an attempt to make thyroxin with too little iodine rapidly recovers.

The same simple remedy has made it possible to rear sheep in the American state of Michigan, where, owing to lack of iodine in the soil and water, they were formerly unable to live. On the other hand, healthy people may make too much thyroxin if given excess amounts of iodides.

Another gland with important internal secretions is the pituitary, which produces at least two different substances.

The posterior part produces a hormone, pituitrin, which affects smooth muscle. It seems to be needed for the normal function of the capillaries, which become relaxed and leaky in its absence, while the kidney produces large amounts of a watery urine. It also has a powerful effect on the womb, and is used to help it to contract during child-birth. The anterior part of the pituitary is concerned in regulating growth. When it is removed we get a dwarfish condition, and when it is enlarged and secretes too much we get overgrowth. The precise effects depend on age. In a child, a long bone, such as the femur, consists of three parts, the shaft, and two bony pads at each end, called the epiphyses.

Growth takes place not at the joints but at the soft cartilaginous junction between the shaft and the epiphyses. If the pituitary begins to over-secrete before the epiphyses have been joined by bone to the shafts, the patient becomes a giant; if afterwards, he becomes a little taller, but the epiphyses, lower jaw, brows, and some other bones grow in thickness, and a condition called acromegaly results, characterized by large hands and feet and a peculiar facial expression. Clearly for normal health the pituitary, like other glands, must produce just the right quantity of its internal secretions.

The adrenal glands, which lie just above the kidneys, also consist of two parts. The central part secretes adrenalin, a substance which produces the same general effects as stimulating the sympathetic nerves; for example, a rise of blood pressure, a slowing down of the movements of the gut, and a pouring of sugar into the blood.

Adrenalin seems to be constantly entering the blood, but in larger quantities during emotion. The outer part (cortex) of the adrenals is necessary for life, and probably pours something into the blood, but we do not know what. The parathyroid glands, which lie in the neck in or near the thyroid, also produce an internal secretion which regulates the amount of calcium in the blood plasma.

The gonads also produce internal secretions, which control the appearance of the secondary sexual characters which distinguish the two sexes. For instance, they act on the nervous system, and cause the appearance of the instincts connected with sex. They also influence the growth of the skeleton and other structures, such as combs and spurs in fowl, horns in deer, and the larynx and beard in man. At the time of writing one of their internal secretions has been obtained in a fairly active condition, though not yet pure.

If we take a few living cells from an organ, and grow them

under the best possible conditions, usually in plasma to which have been added extracts of embryonic tissue, they will continue to divide indefinitely if they are protected from bacteria.

Cells from a chicken's heart have grown in this way for over ten years, a few being transferred to fresh fluid every two or three days. It appears probable that they could be grown indefinitely in this way, so that in the body their death with the natural death of the fowl by old age must be due to causes outside themselves. Blood contains substances which encourage cellular growth, and others which check it. As an animal grows up the former decrease in quantity.

One cell may produce a substance which stimulates another type of cell. Thus a healing wound is full of leucocytes, which crawl into it from the blood-vessels. As these die they liberate a stuff which causes another kind of cell, the fibroblast (which produces connective tissue), to grow and multiply, as can easily be shown on a tissue-culture. The fibroblasts in tissue near the wound therefore grow out into it, forming tough fibrous scar-tissue.

Sometimes growth-regulation breaks down, and the cells of some part of the body grow too quickly, causing a tumour. This may be harmless, like a wart, but if the growing cells migrate into the surrounding tissues and finally to distant parts of the body, the growth, which is then called a malignant tumour, or cancer, is very deadly unless it is removed when still small. We do not yet know exactly why cells become cancerous, though in some cases this condition is undoubtedly caused by chronic irritation, for instance, by tar in tarworkers, by certain parasitic worms in rats.

One of the most important functions of the blood is to clot. If this does not happen, as in certain diseases, a man may bleed to death from a tiny scratch. Blood does not clot in the blood-vessels, even if they are removed from the body, and clots only very slowly in well-greased receptacles. But contact with broken cells or with most foreign bodies leads to clotting; in the latter case apparently because they cause certain of the formed elements in the blood to burst. At least four substances are concerned in clotting. The clot is mostly made from fibrinogen, a protein dissolved in the plasma, but at least three other substances play a part, namely, calcium (lime salts), a protein called serozyme, and a waxy substance called cytozyme produced by the breaking-up of cells.

It is still uncertain exactly how they interact, but interference with any of them will stop the blood from clotting. For example, a little sodium or potassium oxalate will precipitate all the calcium of blood as calcium oxalate, and thus keep the blood fluid.

Of the other proteins in the plasma some are concerned in the defence of the body from infection, though it is possible that fat-like substances play a part too. Infectious diseases are caused by animals and plants, some of microscopic size, some, such as that which causes most common 'colds', too small to be seen with a microscope. One of the best known of the visible ones is the bacillus which causes diphtheria. It grows in the throat, which it may occasionally obstruct, and does not usually spread to other parts of the body, but liberates into the blood a poisonous protein which is particularly dangerous to the heart.

If a man has had diphtheria this protein, called diphtheria toxin, is no longer poisonous to him, and what is more, if some of his blood is mixed with the toxin, it renders it harmless when injected into some one else. In practice one injects ground-up dead diphtheria bacilli into a horse so that it develops antitoxin without having had a sore throat.

Its blood now contains antitoxin and will protect human beings against the toxin. We do not yet know the exact chemical nature of the toxin and antitoxin, but we know a great deal about them—for example, in what liquids they are soluble and at what temperatures destroyed. We also know that the antitoxin puts the toxin out of action by forming a rather loose compound with it. Unfortunately, few bacteria kill us in the rather simple way employed by the diphtheria bacillus, so few diseases can be cured by antitoxins, although they are quite effective against snake-bite.

Another type of immunity is that developed to typhoid

bacilli. The serum of a man who has had typhoid, when added to a suspension of typhoid bacilli in water or salt solution, makes them stick together in little clumps and cease to move about.

The same or similar substances increase the capacity of the white corpuscles for digesting the bacteria, while others will actually cause foreign cells to break up. These substances are mostly very specific, and a serum which is fatal to one race of bacteria will not always kill a very similar race which produces a disease of the same kind, still less those which cause a different disease. Besides such 'immune bodies' in the blood serum which can be transferred from one man or animal to another and experimented with in tubes, the cells of organs can develop immunity which, of course, cannot be transferred, and as to whose nature rather little is known.

In many diseases most of the symptoms are due to changes in the blood-composition, which affect organs remote from the one which has first been injured. For example, in heart disease the first symptom is often faintness after exercise. This is because the damaged heart cannot pump blood fast enough through the lungs to absorb the oxygen needed by the body, so the brain does not get enough oxygen. Many of the symptoms of lung-diseases such as pneumonia are also due to oxygen-want. When the kidneys are diseased they cannot get rid of substances with their usual ease, and they also let the proteins of the plasma leak out into the urine.

Some victims of kidney-disease get headaches, vomiting, and convulsions as the result of the accumulation in the blood of various poisonous substances which a healthy kidney can get rid of, others swell up and get dropsy because the kidney cannot get rid of water and salts. In most acute diseases the blood seems to contain poisonous bodies of unknown nature which upset the temperature-regulating centre, and so cause fever.

Since bacteria and other living producers of disease were discovered we have been enabled to prevent many infectious diseases. For example, we prevent the spread of typhoid and cholera by seeing that their bacilli do not get into drinking-water. Malarial fever is spread by mosquitoes, which suck up from the blood the protozoa that cause it, become infected themselves, and infect the next man they bite by means of their saliva, the protozoa having multiplied and migrated into the salivary glands.

Malaria can thus be prevented by killing off the mosquitoes. But, apart from surgery, rest, good diet, and nursing, we can often deal with disease by means of drugs. Among the most important drugs are antiseptics, which kill off bacteria in infected wounds or abscesses, and anaesthetics, which allow the surgeon to remove diseased organs painlessly. Some drugs, such as Epsom Salts, act in the gut, but most of them have to be absorbed into the blood before they act.

One group of drugs are valuable because they are more poisonous to parasites than to human beings. For example, a dose of quinine which will kill almost all the malarial parasites in the bloodstream only gives the patient a slight headache. But most drugs are valuable because they act on some special organ. For example, there is a form of heart-disease in which the auricles, instead of beating, twitch in an irregular way. The ventricles are constantly being excited by the auricles; they therefore contract so often that they have no time to fill up between beats, and become fatigued.

We give the patient digitalis, a drug which acts mainly on the heart, and in some way blocks the conduction of impulses from the auricles to the ventricles. The latter then begin to beat at a slow rhythm, but fill up properly between beats.

Again, in chloroform or ether anaesthesia there is a danger of the heart stopping, so we want to block the passage of inhibitory impulses from the brain down the vagus. For this purpose we inject a little atropine. This has the property of stopping the action of the parasympathetic but not of other parts of the nervous system. It not only hinders the action of the vagus on the heart, but also on the stomach, and this discourages vomiting.

Many drugs act especially on the brain. Some affect those parts which are concerned in consciousness, and produce sleep or wakefulness as the case may be. Others affect specific centres, such as the heat-regulating or vomiting centres, and by a suitable use of them we can allay pain, fever, and other symptoms which involve the activity of the brain. It is, however, certain that many more chemical weapons in the war against disease remain to be discovered.

To sum up, not only the normal activities of the body, but even its normal shape, depend on the simultaneous normal activities of its various parts. They influence one another partly through the nervous system and partly by mechanical means, as when a bone grows in response to the pull of a muscle, but very largely by chemical factors. The chemical side of the relationship between the hundred million million or so cells that make up your or my body is sufficiently complicated, but it is simple compared to the chemistry of the reactions between the hundred million million molecules that make up a cell of average size.

The study of those reactions is the main task of biochemistry. Suffice it to say that we have begun to isolate many of the intermediate products of metabolism and the catalysts that govern the course of the reactions by which they are formed. But the life of every individual cell is in its way as complex as that of the whole body, in the description of which we have taken the individual cells for granted.

Chapter 8

Physiology of Development

The existence of the fully grown animal or man has been taken for granted, and the attempt has been made to discover as much as possible of its way of working: we have been studying the physiology of adult life. But the adult animal (and plant, for that matter) is not a ready-made article—it has to develop. But that gave us no more than does a mere description of structure for the adult organism. We want to know something of the physiology of development just as much as of the physiology of adult life.

The great difference between the two branches of physiology is this—that while adult physiology is dealing with stability, developmental physiology is dealing with change. In the former case, the general characters of the organism, in spite of alterations in detail, remain the same over long periods of time; and one of the most prominent features of the processes of adult life is that, as a whole, they constitute a self-regulating system, in which any excess in one direction automatically brings into play forces acting in the opposite direction.

But in developmental physiology, the organism is always passing from one phase into another, and even if some of the phases—like the tadpole stage of the frog—are of comparatively long duration, yet processes are always at work which at length will upset the self-regulative power of the system, and will make it change into something quite different. Besides these processes of normal development, however, others of the same general type may occur as the result of special circumstances. The facts of regeneration, for instance, come under this head; and, as a matter of fact, some analysis of regeneration.

Regeneration is not a very familiar fact to most of us, since it is present to a very slight degree in ourselves and in the animals we know best. But worms or polyps, or indeed most of the lower animals—if they could think—would find regeneration the natural, normal state of affairs, and the absence of regeneration a sad and abnormal failure. In a sense they would be right. Regeneration does at least seem to be a fundamental and original property of life, lost late and for special reasons.

If you throw crumpled paper into a freshwater ditch, or set traps consisting of a bottle with boiled earth-worm for bait, you will probably succeed in catching large numbers of various kinds of freshwater Planarians—little leaf-shaped creatures of carnivorous habits and gliding movements.

Choose out the brown and black species, which are the hardiest. Transfer one to moist blotting-paper, and with a sharp knife cut it clean across. The experiment sounds cruel; but in reality you will have artificially encouraged the multiplication of the species.

Each of the bits, replaced in water, will regenerate what is lost and so become transformed into a complete worm. The same would have happened if you had cut the animal into half a dozen cross-pieces instead of two, and equally whether you cut it down the middle or across.

If you persisted, you would discover two main facts in the course of your experiments—that any piece of any shape, provided it was above a certain quite small size, is capable of regenerating into a whole animal; and that this happens equally whether the regenerating process are fed or no.

The same would have been true if instead of a planarian you had cut up the little polyp Hydra, and almost the same with the smaller Annelid worms such as Lumbriculus. The same also holds for the microscopic single-celled Protozoa, although here the operations are much more difficult owing to the animals' small size.

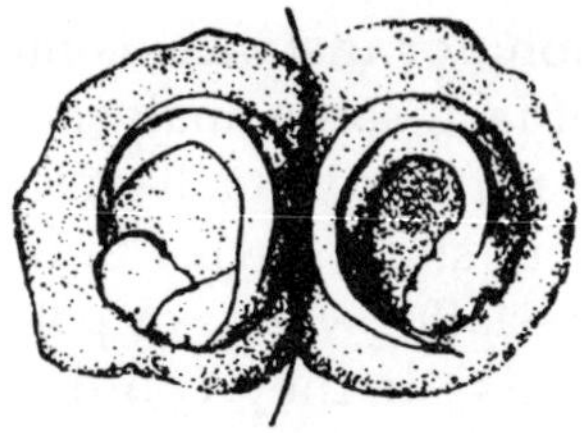

D. Result of complete separation of the first two blastomeres by a hair, *sf*. Two healthy embryos, complete except in size, have been produced.

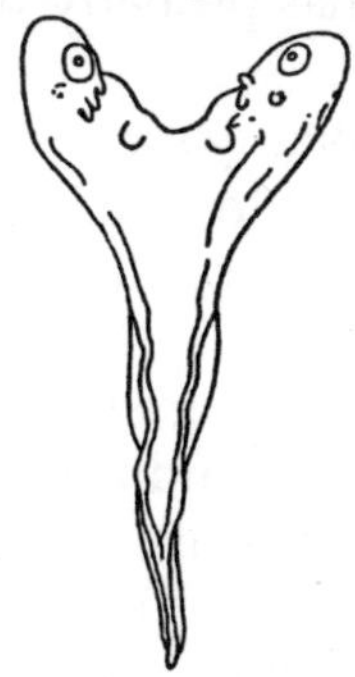

C. the same just before hatching. A two-headed monster has been produced.

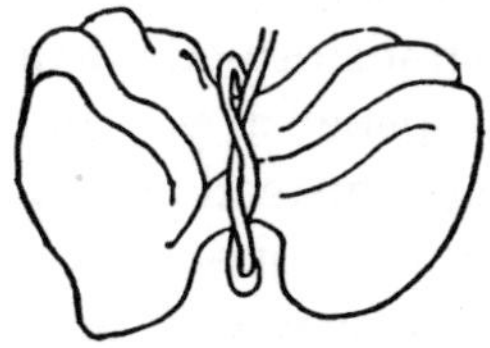

B, Neural-fold stage of the same embryo. Two sets of neural folds have been produced anteriorly; they join posteriorly at *x*.

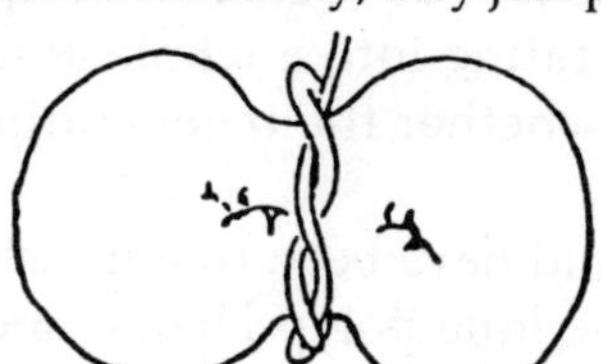

A, gastrula stage after a hair has been tied round the segmenting egg so as to produce light constriction.

Fig. Partial and Complete Twinning Artificially Produced In Newts

It is also true for the very early stages of development in many kinds of animal. In a newt either of the two first cells produced in segmentation may, if artificially separated, give rise to a whole embryo; the same holds in many animals for any one of the first four cells, or for most fragments of the late segmentation and blastula stages.

Thus regeneration of the extreme type—what we may call complete regenerative capacity—is found in the simplest types of animal and the earliest stages of development. It must therefore be thought of, not as a special property developed to meet the rare contingency of losing a limb or being bitten in two, but as something natural to living things, and present unless circumstances forbid it. Further facts confirm this view. In animals of a rather higher grade of organization, such as the more complicated worms, crustacea, some insects, and some molluscs, the power to regenerate is still present, but it is limited. No longer can the animal be cut in half, or into a number of little bits, without losing its capacity for life; but so long as certain central and essential organs remain at work, very considerable losses, such as those of a limb or a tail, can be replaced. This degree of regenerative power continues even into the land vertebrates. Although the frogs and toads do not, the lower or tailed amphibia still possess it.

Even a mature newt or axolotl can grow not merely a new toe or new foot, but can repair the loss of a whole hind limb and limb-girdle. This we may call organ-regeneration. In such animals, the early stages in development of an organ such as a limb are capable of complete regeneration. Thus, by cutting a tadpole's limb-buds so that a number of separated bits are left, the animal may be made to grow more than the normal complement of legs. The same limitation of regeneration is seen, as might be expected, during development also.

For instance, while the amphibian in its segmentation and blastula stages has an almost complete regenerative capacity, the tadpole can only regenerate single organs, and the frog can do no more than heal wounds. Even in newts and salamanders regeneration is much more rapid in the tadpole than in the adult.

Among reptilia, lizards are the only animals which possess even the power of organ-regeneration, and in them it is confined to the tail. When we look closely we find that here the original general power of regeneration inherent in living things has been retained in this organ for particular biological reasons, and has been combined with special adaptations to make it of greater value to the animal. The lizard has the power of self-mutilation (or *autotomy* as it is often called).

If you catch a common lizard by the tail you will find that after a moment or two of squirming, the tail, broken off short at the root, will be all that you have in your hand, and the rest of the lizard will be running off into safe hiding.

This is only made possible by a special structure of the vertebrae of the base of the tail; they have a crack on either side which penetrates almost to their centre, and the tailmuscles are so arranged that when one set of them contracts they pull at the further half of one of these vertebrae, and snap it off across the plane of weakness. The tail when broken off is, though doomed to death within a few hours, not yet dead; and it continues to execute the most violent wrigglings and jerkings for some time.

Fig. Ventral View of a Metamorphosing Toad (*Pelobates*) with six legs. This condition was brought about by cutting into the early limb-buds, each separate piece developing independently.

In Nature what occurs is apparently as follows.—When a bird or other enemy makes a pounce at the lizard, it will often only succeed in catching it by the tail. When this happens, the tail is broken off, and by its squirmings continues to keep the attention of the attacker while the lizard itself is making for safety. The tail is lost and devoured, but the lizard grows a new one. If lizard-eating animals were rational they would probably confine themselves to tails, gathering the annual crop as we gather the annual crops of fruit off a fruit-tree. In any case, the lizard often saves itself by sacrificing its tail (*how* often could be established by finding out how long tail-regeneration takes, and then by examining a number of lizards from one locality and noting how many were in process of growing new tails—an experiment well worth trying); and this desirable end has only become possible by adding the special mechanism for breaking off the tail to the primitive power for regeneration.

Crabs and many other Arthropods have this power of autotomy in their limbs, and some worms when handled break their whole body into fragments, each one of which will regenerate into a complete individual!

One word about the power of regeneration in mammals and ourselves. Although we cannot grow new limbs, yet our wounds will heal, and this is still a real regenerative process; we may call it wound-healing or tissue-regeneration. It is also to be remembered that in a certain sense the growth of such tissues as the epidermis is a continual regeneration—for something is continually being lost, and as continually being renewed.

These last processes fall under the head of *normal* or *physiological regeneration*. A more striking example, perhaps, than the skin is to be seen in the antlers of deer—two masses of bone weighing several pounds which are shed and renewed each year.

Two or three general principles emerge from these facts. The first—to recapitulate—is that regeneration is a primitive property of life. Most low organisms never stop growing, and the capacity for regeneration is clearly associated with the capacity for growth, since growth is necessary for many of the processes of regeneration.

Other changes that are at work in regeneration, however, have the character of rearrangements; for instance, in the formation of a whole newt out of one of the first two blastomeres of the newt's egg, the only process is one of regulation—of rearrangement of material which ought to have produced a half, so that it actually produces a whole of half size. In the regeneration of a small piece of Planarian to form a whole worm, a certain amount of new growth takes place at the cut ends of the piece, but many of the changes involved are brought about by changes in the original piece. For this to happen, however, a certain degree of plasticity is needful—cells and tissues must be capable of altering their original character and adapting themselves to new conditions.

This also is largely a property of actively growing organs; further, it appears to be impossible if tissue-differentiation has reached too high a pitch. Thus the gradual limiting of regenerative capacity which we find as a general rule both in the development of the individual and in the progress of life as a whole, is due partly to increased differentiation, partly to the fact that the higher animals, instead of growing indefinitely, show a sharply marked stoppage of growth on reaching the adult state.

Regeneration is thus essentially the restoring of typical form, structure, and function; we might call it a special case of regulation. The fact that a piece of Planarian without a mouth can yet regenerate head and tail is obviously interesting. The piece's capital can be readily transformed into cash for the undertaking of new enterprises—in other words, the living framework of its body can be broken down into foodmaterials which can then be used up in new growth elsewhere. In the higher animals, the living capital is locked up in non-negotiable forms to a much greater extent, although not so completely as is often imagined.

This power of drawing upon the living tissues in case of need leads on to another remarkable power of Planarian and other low types. If an intact animal of this sort is starved, it can continue to exist for periods of weeks and months by thus liquidating its capital for its day-by-day expenses. Since it draws on its own tissues for its daily energy-needs, it therefore must grow smaller and smaller. Planarians can continue this economical process until they have diminished their size to below that at which they hatched from the egg. If they are given food again (at any stage in the pro-ceedings) they will start growing once more, and recover their normal size and appearance. One very curious fact was early noted about Planarians thus reduced in size by starvation—namely, that their proportions gradually altered from those of the adult and became more like those of normal young worms. This led to the idea that perhaps they not only *looked* young, but had really *become* young again—an idea which was easily put to experimental proof.

Some species of Planarians reproduce mainly asexually by transverse division; after the worm has reached a certain length a new head is formed a little behind the centre of the body, and in front of this a split appears along which separation eventually takes place. A brood of worms from one such species was taken. When they had nearly reached full size, they were divided into two lots.

One lot, fed regularly, continued to grow and divide in normal fashion. The other lot was starved until the average size of its members was reduced to about half; then it was fed until it had recovered its original size; then starved again—and so on repeatedly. The experiment was continued for a period of time in which the well-fed animals went through nineteen generations (a period which, though only occupying a few months for Planarians, would in human beings mean about six centuries —from Chaucer's time till to-day). During the whole of this period, none of the other lot reproduced at all, but all remained within the limits of size set for them by the experimenter, and, what is more, showed no signs whatever of age or of diminished vigour.

In fact there is every reason to suppose that the experiment could have been continued indefinitely—in other words, that the individual Planarian, by proper treatment, can be made immortal. The elixir of life was sought by the alchemists of medieval Europe for hundreds of years. These experiments show that it has at last been discovered; but, unfortunately, it is only effective with Planarians and some other animals, all of a low grade of organization. Before we go further into the principles involved in this matter, we must mention that there is another way in which animals may surmount starvation and other unfavourable conditions.

If an ordinary hydroid polyp like Obelia be kept in the laboratory, the hydranths, or separate organized individuals of the colony, will (unless the water is well aerated artificially) show a curious series of changes.

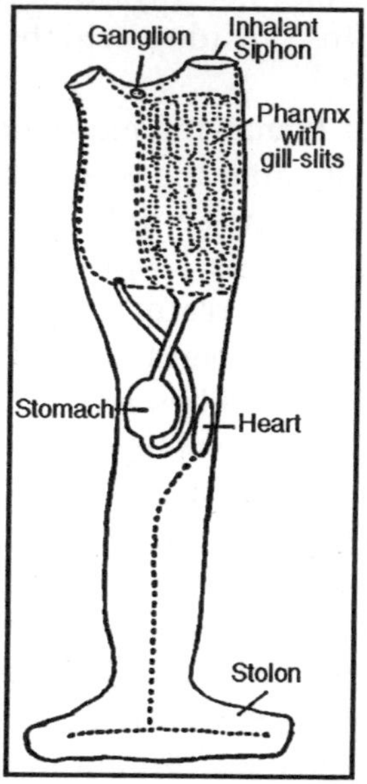

1, normal.

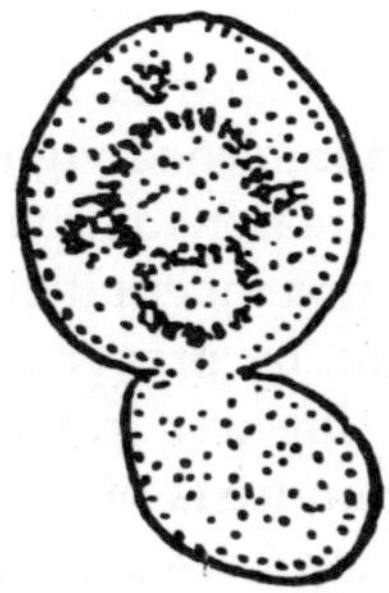

2, the same Individual Dedifferentiated after exposure to adverse conditions.

3 the Flatworm, Planaria. 3, Normal adult.

4, the same Individual after several months without food.

Fig. Dedifferentiation and Reduction

Their mouths will close and their tentacles become stumpier; eventually they will lose almost all signs of their previous differentiated structure and become converted into a rounded or egg-shaped bag with a few little knobs marking where the tentacles had once been. They have lost differentiation—in other words, have undergone dedifferentiation.

In Obelia the process is complicated by the fact that while differentiation is going on, the cells of which the hydranth is made up detach themselves from their neighbours and, leaving their fixed place in the tissues, migrate into the central cavity and down into the common stalk which connects all the hydranths. By this means the hydranth dwindles rapidly, and is reduced soon after dedifferentiation is complete, to a tiny knob in the bottom of its protective horny cup. The animal has been resorbed by the stem. It is as if a house were to be unbuilt, by the bricks leaving their places in the walls and migrating into the garden.

If, however, the hydranth is cut off at its base beforehand, the central cavity is soon filled up with detached cells, and after this, dedifferentiation alone occurs until the animal is reduced to a mere living bag tight packed with cells. These types of dedifferentiation are found in many low types of animals, such as various Coelenterates and Ascidians.

Among higher forms partial dedifferentiation may be a normal mode of development. For example, when the tadpole metamorphoses into the frog, some of its tissues start to dedifferentiate, for instance, those of the gills and tail. The process is complicated by the fact that phagocytosis, or the devouring of undesirable materials by white blood-cells, here reaches a far greater pitch of perfection than in lower animals, and, as a result, as soon as the cells have reached a certain degree of dedifferentiation, they are attacked and removed by the white blood-cells. The disappearance of the tadpole's tail is thus due more to the activity of the phagocytes than to the migration of the tail-cells themselves. Similar methods of resorption of tissues are found when, for instance, a tumour or a graft disappears in a human being.

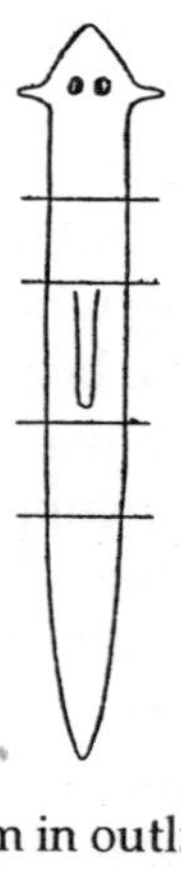

1, a normal adult worm in outline, showing eyes and pharynx; *a* and *b*, levels where pieces were cut out.

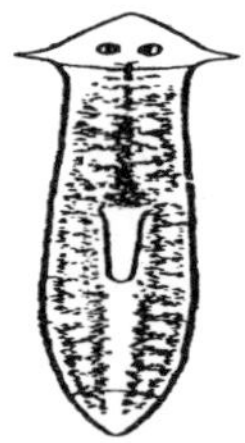

2, either *a* or *b* pieces may regenerate a new head; in this case complete differentiation to a normal worm occurs. The original pieces produce new head and tail by new outgrowths, new pharynx and altered shape by remodelling.

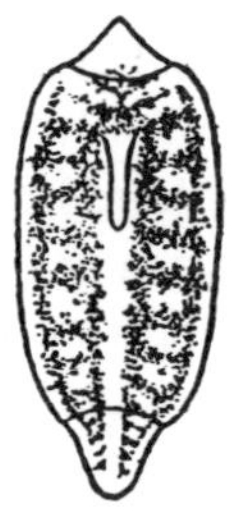

3, in adverse conditions, only a rudimentary head is produced.

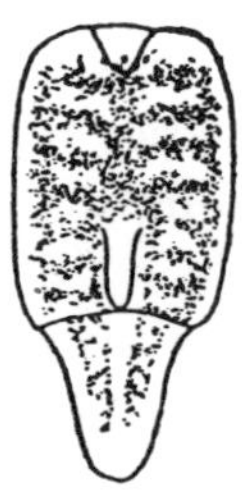

4 and 5, in very adverse conditions, no head is formed; the front end merely forms a scar. In these circumstances, *a*-pieces form a new pharynx (4), but *b*-pieces do not (5).

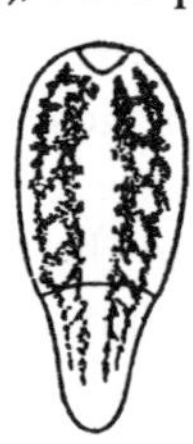

Fig. Regeneration and Differentiation in Planaria

To return to regeneration, and to another of its aspects, it is found that the presence of one part is often necessary for the regeneration of another part. In Planarians, for instance, the mouth opens at the end of a sort of trunk, the muscular pharynx, which can be protruded from the centre of the underside.

If a small piece of the flatworm is cut out by two transverse cuts, it depends on a number of circumstances whether a head shall be formed or not; for instance, a low temperature near freezing, or exposure to dilute alcohol or other poisons (the experiment is easy to carry out) will reduce the frequency of head-formation, or even suppress it altogether.

If a head is produced, such a piece will always proceed to the formation of a pharynx. But if no head appears, the formation of a pharynx depends on the original position of the piece in the body. If it was cut from in front of the original pharynx, it will often regenerate a pharynx, even in the absence of a head; but if it came from behind the original pharynx, a new pharynx will never be formed unless a head forms first.

In other words, a new pharynx will be regenerated in the centre of a piece of Planarian provided that there exists at the front end of the piece a region which is normally anterior to a pharynx—whether this anterior region be an old part of the body or a new regenerated head.

Thus, the head or front region of the body exerts some remarkable influence upon a distant region of the rest of the piece, making it construct an organ that it could not otherwise produce. It makes things make themselves. Exactly how this effect is produced we do not know. Possibly it is through the agency of the nervous system. In any event, it is clearly a property of very great importance, and also one which seems to be of quite general occurrence. For instance, when the eye-stalk of a prawn is cut off, a new one like the old is usually regenerated.

But when, in addition to the eye-stalk being amputated, the optic ganglion, or part of the brain to which run the nerves from the eye, is also cut out, the organ which is regenerated bears no resemblance to an eye and eye-stalk, but has precisely the form of part or all of the first antenna or feeler, the appendage which normally comes next behind the eye. One other fact gained from the study of regeneration needs to be considered before the problems of normal development. If one of a pair of organs be removed, remarkable results often follow.

For instance, in certain diseases, one kidney may have to be cut out. When this is done, the other begins to grow, and at the end of a few months weighs almost as much as the original pair together. This enlargement follows as a direct result of the increased demands made upon the functions of the surviving kidney, and is therefore known as compensatory functional overgrowth. There is no regeneration of the missing organ, but a compensatory growth of its mate, resulting, one may say, in a regeneration of function.

This functional overgrowth may be found not only to compensate for the loss of one member of a pair of organs, but also when part of a single organ is removed, or when an extra demand is made upon an organ without removal of anything. If, for instance, most of the liver or pancreas or reproductive organs are removed, the bits that are left will start to grow rapidly, until they are large enough to cope with the demands of the body.

Or, as illustrative of the second case, the normal thyroid secretion is very rich in iodine. When the amount of iodine in the food and drink falls below a certain quantity, the demands of the body for more iodine-containing secretion react on the thyroid; this begins to grow, producing more total amount of secretion as an attempt at compensation for its poverty in iodine. It may grow to a relatively huge size, producing a great swelling, called a goitre, in the neck. If a little iodine is given to a patient with such a goitre, the demand ceases, the reverse process sets in, and the swelling is reduced.

A similar process takes place in muscles when extra demands are made upon them. Everybody knows how heavy exercise 'develops the muscles', and, after a preliminary period of loss of weight due to utilization of the reserves of fat, makes a man put on flesh. The actual size of the muscles increases as the result of exercise. With this information we can now attempt some account of the normal development. It is found that the early development of a vertebrate such as a frog or newt falls, from our present standpoint, into three main periods. The first is essentially a period of the division and rearrangement of already existing materials, and takes the fertilized egg through segmentation and some way into germ-layer formation.

The second is mainly one of differentiation—visible structure appears, the organs are blocked out, and tissues become different each from the other. Finally, in the third, the two most important features are growth, and the welding of the differentiated parts into a working whole—by mutual influence of organs upon each other and by an adaptation of separate organs to the demands upon them through the effects of function upon growth. This last phase may be taken first, since it is in many ways the most familiar. Think of a young organism—a recently hatched tadpole, a puppy, a child. Although it has already a characteristic organization, yet this organization is still plastic within wide limits. The final shape and size of its bones, its muscles, its sinews and tendons, its glands, its digestive tube, even of parts of its central nervous system, depend very largely upon the amount and the kind of use to which they are put during development.

If one of a puppy's fore-limbs is tied up soon after birth so that it is never used, the dog will manage to get about very well with only three legs. But the leg which has not been used will be very different from the others.

A

A. Six years old, immediately after the operation; the grafted piece of bone (x) is of an irregular shape.

B

B. Two years later (the position of the hand is reversed), the grafted piece has become moulded into a very good imitation of the original joint. (After Timann.)

Fig. X-ray Photographs of the Hand of a Boy In whose third finger the Basal Joint became diseased; it was removed and a piece of healthy bone with its bone-forming Membrane (Periost) grafted in from another situation.

The bones of its skeleton will grow almost to the normal length, but will never attain more than about half their normal thickness, nor will the arrangement of bony struts and stays at either end of the shaft (which in an ordinary bone are developed so as to meet both vertical and sideways strain in the mechanically most perfect way) be properly formed. The muscles of the leg, too, and their tendons will be very small. Not only this, but the nerve-cells of that region of the brain which controls the movements of the limb will be undersized.

The power of bone to respond to the strains to which it is exposed, is of practical value. If, for instance, one of the small bones or phalanges of a finger is crushed and has to be removed, a small piece of bone from some other region can (if it include some of the periost layer from which new bone-cells are produced) be grafted in to take its place. After a year or two the irregular splinter will have come to bear a marked, though not complete, resemblance to the bone for which it is doing duty.

One of the most striking adaptations in the body is the fact that every tendon which attaches a muscle to a bone not only is of the proper size for the strain which the muscle can exert upon it, but is composed of parallel fibres which run precisely in the region of greatest stress. How can this beautiful relation of mechanism to function be supposed to have originated by natural causes? Does not the purposeful adaptation, different in detail for each separate tendon, demand the intervention of some supernatural power?

Research, however, has in great measure removed such difficulties; for it has shown that all the separate detailed adaptations are the direct consequence of one inherent property of the tissue of which tendons are made. If a piece is cut out from a long tendon, such as the Achilles tendon which passes from the muscle of the calf over the heel to be fastened to the sole of the foot, it will after a time regenerate.

Connective tissue grows out into the wound, the cells elongate and form fibres in the direction of the tendon, make attachment to the cut ends, and reproduce a structure essentially like that which was originally present. But if by one means or another the calf-muscle is put out of action (as when its nerve has been severed) then, although the gap will be filled by a new growth of connective tissue, the fibres will not become arranged in any particular direction, and no tendon-like structure will be formed.

Further, when in such an animal (the main experiments were carried out in rabbits) a piece of clean silk was healed across the middle of the wound, at right angles to the original tendon, and then, by winding it up very slowly on a screw, increasing tension was put on one end of it, it was found that fibres were formed in the new connective tissue, but parallel to the silk, and therefore useless for any restoration of normal function.

From these facts (supported by many others) it is clear that tendon-fibres are formed from connective tissue when the connective tissue is frequently stretched, and further, that they are formed parallel to the stretching force. This stretching force is usually the force of muscular contraction (in growing animals there will also be found stretching due to growth of bones); so that tendons will be formed from ordinary connective tissue when muscles begin to contract, and will be 'adapted' to the direction and amount of stretching exerted by the muscle. In any case, all the apparent design seen in the adaptation of each several tendon to the demands made upon it, is apparently a direct result of this one reaction of connective tissue to tension.

To sum up, we may say that certainly most of the tissues of the body—possibly all—respond either throughout life or at least during growth (while their cells are still capable of multiplication) to the demands made upon them, and respond on the whole very advantageously from the point of view of the whole animal. The skeleton, the muscles, tendons, many and probably all glands, the central nervous system—all these we have already seen to show functional adaptation. So does the skin.

Everybody knows that (after the blister stage has been negotiated) continual rubbing will make the skin thicken. 'Horny-handed' is so hackneyed an epithet for manual labourers as to have almost become mere journalese, and every oarsman demonstrates the fact in his own person. So, too, the circulatory system responds to functional demands.

If Everest is ever climbed, it will probably be because men can become 'acclimatized' to high altitudes and consequent low oxygen supply, largely, it would seem, through an extra production of red bloodcorpuscles by the bone-marrow and in consequence a greater capacity of the blood for capturing and transporting what oxygen there is.

Furthermore, the size of a blood-vessel, both diameter of cavity and thickness of wall, is determined very largely-perhaps wholly—by the amount of blood which is pumped through it, and the pressure at which it is pumped. Cases are known in which large arteries become blocked up.

If the blocking-up process is not too rapid, one or more small branches of the vessel will gradually enlarge, and produce channels capable of transporting the requisite amount of blood without difficulty. This has been known to happen even when it is the human aorta, the main and indeed the only large artery springing from the arterial side of the heart, which has been blocked.

The same kind of adaptation is also true of the protective

reactions of immunity. It is a familiar enough fact that children who have had measles once do not generally catch it again, but this is only one among innumerable instances, such as the over-production of diphtheria antitoxin by horses injected with small doses of the toxin or poison produced by diphtheria germs, which has reduced the disease from a cruel scourge to a serious unpleasantness, or the benefits given by vaccination.

All these again depend on one single property of higher animals; when unaccustomed proteins enter their system, they can (if the foreign substances are not in too great amount, and provided certain other conditions are fulfilled) not only destroy or inactivate them, but in the process generate an excess of the destroying substance or *antibody*, which remains for a longer or shorter time so that a second invasion can be more rapidly and effectively crushed.

Although many of these adaptations are possible throughout life, others can only occur during development, and most of them are essential for normal development, since only through them do the size and structure of the various organs come to correspond to their function and to the physical and chemical stresses imposed on them.

There is, however, another type of mechanism which is active through the later periods of development and plays an important role not only in this functional adaptation but also in fusing the parts of the organism into a more unified whole, and that is the endocrine system or sum total of the ductless glands.

Many of them, such as the thyroid and the pituitary, have indispensable functions as regards proper growth. The thyroid regulates the rate of metabolism, increased amount of thyroid secretion producing an increase of oxidation in the body. As might be expected, this function seems to be of more importance to the well-being of animals such as mammals, with a normal high metabolism (great heatproduction and constant high temperature), than to those like fish or frogs, which do not have to be constantly generating large quantities of heat to keep their temperature above that of their surroundings.

Within the mammals, the absence or failure of the thyroid makes more difference to an adult man than to an adult sheep: the thyroidless sheep is but slightly abnormal, while the thyroidless man becomes notably fat, sluggish, and slow-minded. But its absence in the child, in whom growth and activity should be much greater, makes far more difference again than in the grown-up man; the thyroidless human child or *crétin* usually dies early, can only grow into a stunted dwarf, and never develops normal intelligence, the brain of man apparently requiring a high level of metabolism for its unfolding. In Amphibia, metamorphosis from larva (tadpole) to adult is under the control of the thyroid.

Tadpoles with their thyroid cut out refuse to be transformed into frogs, and grow to giant size in the water, while if a dose of thyroid is given to quite young tadpoles, they will metamorphose into froglets no larger than flies. The changes which take place in body and mind at puberty are also under the control of the endocrine system. The immediate control is exerted by the gonad itself. It is important to find that a whole set of separate changes can be brought about by a single change in the ductless glands, and further, that all these changes are in the long run related to one function, that of reproduction.

That a single original alteration in the chemical composition of the blood, an alteration appearing inevitably at the end of a definite span of growth, should bring about an increased growth-rate, the formation of mature spermatozoa in the testes, the appearance of hair on the face and body, and a marked change in mental outlook and temperament, is one of the most striking facts in biology.

All the ductless glands have this in common, that their secretions are carried all over the body by the blood, so that they can exercise a number of different effects simultaneously upon the most diverse organs. It is fairly obvious, however, that for all the importance of function and of the endocrine system in development, their task is essentially one of regulating the degree of activity of processes already in being, or of directing them along particular channels.

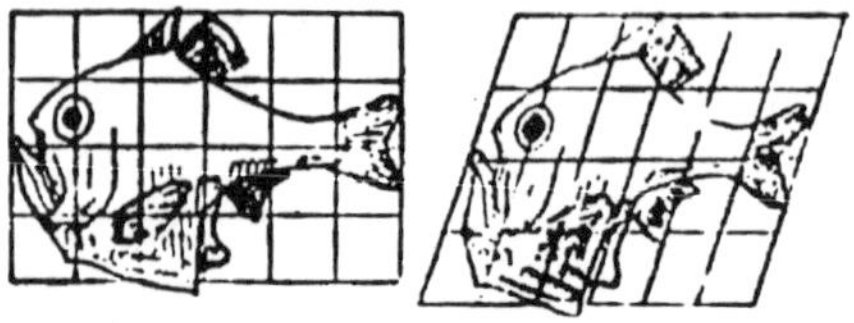

Fig. Outlines of two Related Species of Fish. The one can be derived from the other by a simple distortion of the relative directions and intensities of growth in length and growth in a dorso-ventral direction.

Bones exist in the absence of a pituitary or of pressure upon them, and muscles without being used. It is only a small part of the adult man's growth of hair that is dependent upon the testis; metabolism continues in the absence of a thyroid, and the first development of a gland or blood-vessel takes place before its use begins. Another point that should not be forgotten is that in the long run the differences in shape and size between different animals depend almost entirely upon differences in the amount and direction of growth in different regions during development.

In very early life the hind-limbs of a whale grow much less fast, compared with the body, than do those of a man, and those of a man much less fast than those of a kangaroo. Some idea of the changes which would be produced in human beings if the proportions between general growth in length and growth in breadth were rather different from what they actually are, can be obtained by visiting the distorting mirrors at a fair or circus.

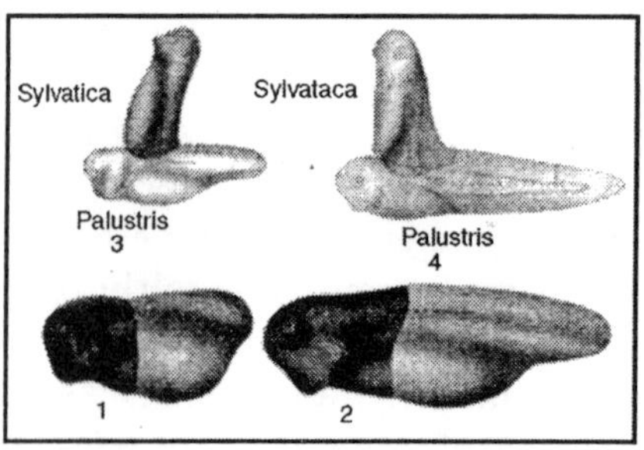

Fig. Two Experiments in Grafting in Tadpoles

Or one organ may be affected differently from the rest; the effect of this on general proportions may be seen by looking at oneself in the broadening mirror, first with the arms held against the sides, and then outspread at right angles to the body.

- The front half of an embryo of one American species of frog (*Rana sylvatica*) has been cut off and grafted on to the hind half of another, lighter-coloured species (*R. palustris*).
- The compound animal (chimaera) grows quite normally. The lateral line, which originates near the head, has grown down from the *sylvatica* component on to the *palustris* trunk. Compound individuals like this have been reared through metamorphosis.
- The front part of a *sylvatica* embryo has been grafted on to the back of a *palustris* embryo: this combination also has continued development. The *sylvatica* lateral line (below *si*), on reaching the *palustris* component, has bent round into correct position.

If in the embryo of the fish Argyropelecus the tendency to growth in length had been a little less, and the main direction of dorso-ventral growth were not quite at right angles to the main direction of growth in length, it would have developed into a passable imitation of another kind of fish. No doubt such slight alterations in growth are at the bottom of much evolutionary change.

How do these more fundamental structures and functions originate to provide the primal basis to be shaped by the secondary moulding forces we have described? Although much is still not clear, yet the main outlines of the answer emerge when we look at the next previous stage of development, the stage of differentiation.

This starts during gastrulation, and ends as the various organs become capable of exercising their functions. Its great characteristic is the sudden appearance, in the hitherto comparatively simple germ, of a number of chemically different regions, each one of which is irrevocably destined to generate one particular kind of organ. What is more, this will usually be accomplished whether the developing organ-rudiment is in its proper relation to the rest of the organism or not. If the late gastrula is cut in two so as to separate front from hind region, each part (after healing the wound) will proceed to develop just those organs which it would have done if the embryo had been left intact. We are presented with an anterior half-embryo without any of the hinder organs, and a posterior half-embryo with tail, hind limb-buds, and most of the trunk, but no head-regions, gills, or heart.

There is, in fact, in this stage *no* regeneration whatever of missing parts; yet, in spite of this, both halves appear quite healthy, and only begin to degenerate at a stage when the organs should be beginning to work and the animal to fend for itself. Or a piece of one embryo may be grafted on to another embryo, even of a different species, and both will continue to develop. One actual 'chimaera' has been artificially produced consisting of the front half of one kind of frog and the posterior half of another, which grew and metamorphosed quite happily.

Fig. An embryo which during its gastrulation has had grafted into its flank a dorsal-lip region from another embryo.

Or if a small part of the future brain-region be cut out just after the gastrula stage, and grafted in again in reversed position, it will produce what it would have produced in an untouched embryo, although now of course the organs to which it gives rise will be misplaced. The eye, for instance, which is formed by an outgrowth from the brain, may by this means be made to develop behind the ear, and yet have normal structure. This will happen in spite of the fact that no visible trace of even the earliest eye-rudiment was present when the operation was performed. Before gastrulation, on the other hand, no such irreversible changes, inevitably determining the future appearance of organs, have yet set in. Cutting the developing egg in half in the earliest stages has a totally different result from cutting it in half after gastrulation has set in. In the former case, as we have seen, one or both halves may become reorganized into a whole, and produce an embryo normal in every respect save size. One very remarkable fact has recently been discovered, namely, that the sudden change that takes place in a few hours during gastrulation in an animal like a frog or newt, during which the germ loses its capacity for regulation and becomes a sort of jig-saw of separate parts, each predetermined to produce just one particular organ of the future tadpole, occurs under the influence of one small region of the embryo.

This is the region just above the dorsal lip of the blastopore. By ingenious microscopic operations, even tiny bits of a newt's egg can be removed and grafted into other places on the same or other embryos. If this dorsal-lip region be taken and grafted into the flank region of another developing egg, it is found to cause the development of an extra set of organs, in addition to the one produced in relation to the host egg's own dorsal lip. An extra nerve-tube, notochord, muscle segments, kidney tubes, and gut may be thus induced to form—in other words, the beginnings of a whole second embryo. What is more, they are not formed by the grafted piece itself, but in the host-tissues near it. Again, we are reminded of Mother Carey in Kingsley *Water Babies*, who 'made things make themselves'. No other region has this power, and so the title of 'organizer' has been bestowed upon the dorsal-lip zone.

It is interesting to recall that a similar sort of power was exerted by the head of a Planarian in inducing the formation of a mouth and pharynx. Another fact is that the dorsal-lip region at this period is the most actively working part of the embryo, with very rapid cell-division. It seems probable that this great activity has something to do with its organizing capacity.

Other interesting results, also apparently concerned with

differences of activity in different parts of the developing egg, have been obtained with still earlier stages.

A. Normal larva, with anteriorly placed mouth (M).

B. Incompletely cyclopean larva, with the two eyes joined and occupying the position usually taken by the mouth.

C. Completely cyclopean larva, with single antero-median eye. Dorsal aspect.

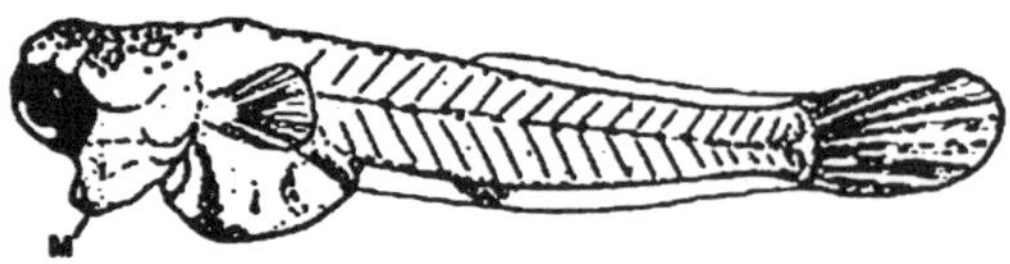

D. Lateral aspect of same, showing the ventral mouth (M).

E. Ventral aspect of same; ys, yolk sac. (From Stockard.)

Fig. Free-swimming Larvae of the Top-minnow

Although the period up till early gastrulation is of course an essential link in the developmental chain, no important steps in the construction of the details of the animal's typical organization are being taken so long as it lasts.

The original comparatively simple organization of the zygote is not much altered; and the most essential change is the cutting up of the huge and unwieldy single cell, the fertilized ovum, into hundreds of small cells. In the first place, these are far more mechanically convenient, just as a hundredweight of bricks is suitable building material, while a single giant brick weighing a hundredweight would be useless.

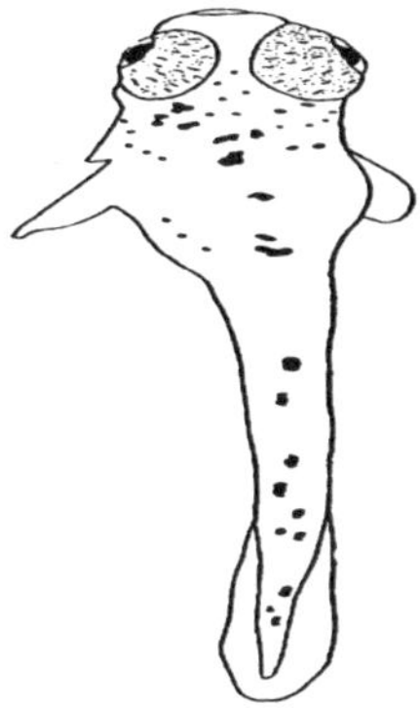

A. Control, developing in normal conditions.

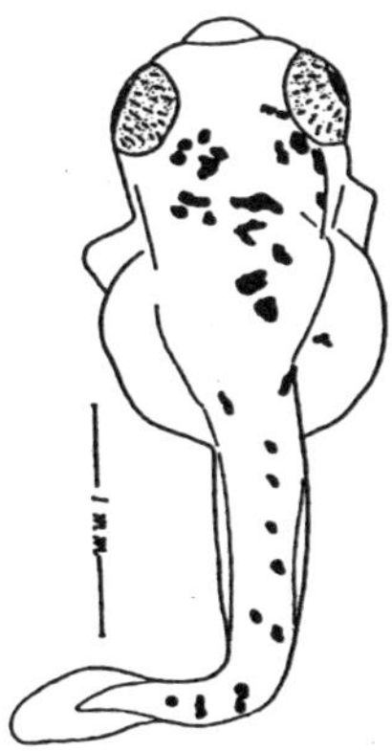

B. Treated specimen. Note the much greater size of the head. Its heart-beat was also more rapid. (After Gowanloch.)

Fig. Acceleration of Development in The Embryo of a Fish

And then each is a relatively independent unit, with its own cell-membrane as frontier, so that it is much easier to localize a number of separate chemical processes in different parts of an embryo divided up into cells than it would be to localize them if the embryo were all one large cell, in the same sort of way that the possession of a number of small pots and pans makes easier the task of preparing an elaborate dinner.

To choose another parallel from human affairs for the second and third phases of development, we might say that the second corresponds to the discovery and invention of new processes, new machines, new products, while the third corresponds to the growth of the industries based on these discoveries, a growth which is regulated by the laws of supply and demand, by the customs and laws of the country, and by the state of world affairs, but can only take place once the new inventions have been made.

In conclusion, it is worth mentioning that every cell of the body appears to receive the same complement of chromosomes, the original set possessed by the fertilized egg being exactly duplicated at each cell division, so that each cell has a 'copy' of it.

This at first sight seems to make differ-entiation hard to understand; but perhaps a musical analogy will make it clearer. The performance of a piece of pianoforte music demands a pianoforte. But we might have a hundred identical pianos in a row, from each of which would be emanating a wholly different piece of music: that would depend upon the players.

For the pieces of music substitute the different organs of the embryo; then the pianos are the chromosome-sets and the players are represented by the different conditions in different regions. For instance, at the close of segmentation some cells are small and poor in yolk, those at the opposite end are large and yolky, some are at the surface, others tucked away in the interior; and the differences between different parts are increased after gastrulation.

In animals progressive development generally comes to a close with the attainment of the adult condition. In man, however, this need not be so in respect of one character. Mental development may continue; in other words, the contents of the mind may increase and its organization become more complex and more perfect, so long as maturity continues and senility has not set in. Of course this is by no means always very marked, and equally of course something of the same kind is seen in the higher vertebrates which can profit by experience. But the extent of the possible development open to man is none the less very striking.

One of the chief biological differences between man and all other organisms is that, through his power of thinking and learning, his development and differentiation can continue through the whole of life.

Chapter 9

The Methods of Evolution

The simplest of actions are often dependent upon a delicacy of mechanism not to be found in our man-made machines, while many human faculties could never exist without the foundation of a whole series of carefully adjusted living processes. Human thought is only possible within narrow limits of temperature. Our brains will only work even approximately well within a range of about 3°or 4°C, and to keep the blood at constant temperature requires the elaborate machinery, of sweat-glands, dilating and contracting blood-vessels, and their nerve-supply.

Or again, merely to pick up a pencil from the floor or a cricket ball from the grass requires first of all an exact judgement of distance, which means a connexion and co-ordination between the brain processes involved in sight and the judging of seen objects, and those involved in the sense of muscular movement and judgements of muscular movement; and secondly, the co-operation of a large number of separate muscles in body, arm, and fingers, each of which has to be made to contract or relax just so much, no more and no less, by nerve-impulses of just the right number and frequency.

How is it possible that this complexity can have come into being? At the outset we must remember that it *does* come into being in every one of us during the first few years of our life. Each one of us started as a formless germ with hardly any differentiation—no separate organs or tissues, no senseorgans, no brain, no capacity for regulating its temperature, for moving from place to place, for thought.

Something (and a great deal too) had to be formed out of

next to nothing before we were ready to be born. Even after birth we had to learn to do a great many things that we take for granted, such as judging the shape of things from sight or even from touch; learning to control our muscles accurately. Have you ever watched a baby making bad shots for its mouth, or at some bright object in front of it, learning to walk, to speak?

A new-born baby is even more helpless in many ways than it seems. Its brain-cells and fibres are not fully developed, and it is not for some weeks that it is even capable of beginning to learn many types of lessons.

The power of the centres in the cerebral cortex to combine the raw materials provided by the senses in proper relation is well seen in the following experiment. A man was provided with inverting lenses, so that he saw everything upside down. By wearing them all day for a week or so, however, he gradually came to make the right associations quite unconsciously between the inverted vision and his bodily movements. It is in a similar way that the baby must learn to associate its visual and tactile and muscular sense-impressions, for they are all equally arbitrary, and only by association can the relation between them be worked out by the brain.

There has thus been real evolution in the individual development of every human being and higher animal or plant—an evolution that is accessible to everybody's observation.

Characters acquired in this way during life are usually called *acquired characters*, or sometimes simply *modifications*. If, for instance, a tame sea-gull is fed on corn instead of fish, the whole lining of its stomach alters, becoming thicker and more like that of seed-eating birds. Or, again, the proper sort of exercise actually causes muscles to become larger. In hard training, although at first a man's weight will go down, because the fat gets used up as fuel, yet afterwards his weight will go up again, this time owing to the building of new muscle-tissue.

The Lamarckian theory would have it that the effects of changes like these are perpetuated in the offspring; thus the direct action of external conditions, or, as we usually say, of the animal's environment, would be one chief cause of evolution, and the habits and will of the animal the other chief cause.

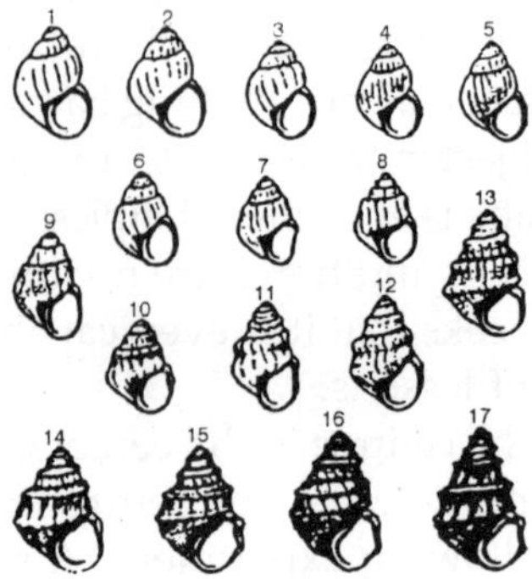

Fig. To Show Intergrading Variation in 17 existing forms of the freshwater snail *Paludina,* from various localities
The extremes would be regarded as distinct species if they were not connected by a complete series of intermediate forms. Similar gradual variation is often found during geological time

However, there is little or no evidence for the theory, and a great deal against it. We will take a particularly striking example from botany. Many plants grow both high up among mountains and down in the plains and valleys. Those which grow higher up must exist in lower temperature, longer winter, more violent weather, poorer soil.

As a result, they grow into a shape quite unlike that of the lowland variety. The Dandelion, for instance, when growing in the Alps is a dense rosette of small leaves, a long root, and a short flower stalk; in the lowlands, as we all know, its stalk may be long, its leaves large and spreading, its root (though too long for the gardener) much shorter.

If Dandelions from the Alps are taken down and grown in the plain, all their new growth is of lowland type, and in a short time they become indistinguishable from lowland dandelions, while exactly the reverse is true of lowland plants transplanted to the Alps. The same is true when seeds of the Alpine type are sown in the plains, and vice versa. The time spent in an Alpine environment has not in the least fixed the

Alpine habit of growth. This can be readily understood if we suppose that the Dandelion has a fixed constitution which, however, reacts differently to different external circumstances; it has, that is to say, a fixed capacity of being modified in special ways.

Fig. Modifications in the Dandelion (Leontodon) Produced By Differences In The Environment

This is obviously true for simple chemical substances. One particular sort of paraffin, for instance, melts at, say 61°, another at 62°C Keep the former in one environment-below 61°—and it stays solid; keep it above 62°, and it stays liquid. But you will not make it raise its melting-point however long you keep it melted. The two will continue to show their characteristic modifications in relation to heat-their original melting-points—as long as you like to keep them.

To take another example, this time from the animal kingdom, attempts have been made to explain the black colour of the negro as the accumulated effect of generations of sun-burn. It is of course true that most white men in a tropical climate become sun-burnt; the strong sunlight has a direct effect in causing more pigment to form in the skin. But anybody can verify for themselves the fact that the children of men tanned in this way are not noticeably less white than those of parents who have never left England.

What does this imply? Surely, that European and negro have different constitutions as regards skin-colour. The

European stays pale in temperate countries, darkens in the tropics. But the negro is black wherever he lives. We can go back to our previous comparison; we can think of two grades of paraffin, one melting at 65°C, the other at 40°C. The first, if kept in the shade, will remain solid throughout the summer in any region of the earth; but the other, though it would stay solid in an English summer, would be liquid in the tropics.

This interpretation would clearly leave no room for the Lamarckian theory, and, as time has gone on, more and more of the examples that the Lamarckians claimed as proving their point have been shown to be explicable in some such way as this.

A, plants, all of similar hereditary constitution, grown under uniform conditions of environment: result, similarity.

B, plants of various hereditary constitution grown under similar conditions of environment: result, diversity, due to utation or to recombination.

C, plants, all of similar hereditary constitution, grown under various conditions of environment: result, diversity, due to modification.

Fig. The Effects of Differences in Environment and in Constitution Upon Differences in Visible Characters of Organisms

The further difficulty remains that we are completely in the dark as to *how* the inheritance of acquired characters could be brought about. Every animal consists of a number of organs concerned with the working of the individual, together with the reproductive cells whose primary function is the perpetuation of the race. To the first, taken together, the term *soma* (the Greek for body) is applied, while the second is called the *germ-plasm*. The soma is of necessity mortal, while the germ-plasm is potentially immortal, and is the continuous chain on which the individuals of the race are strung. All acquired characters—that is to say, all characters which the Lamarckian theory supposes to be accumulated in evolution—are *somatic*; they affect the soma.

But how is a change in the soma to alter the germ-plasm? And even if it did, how is it to alter it in such a way that when the new soma of the offspring is produced, it shall show a change of the same sort? As Weismann, the well-known German zoologist of the late nineteenth century, put it, we might equally well suppose that a message sent off by telegram in Paris would arrive in Pekin already automatically translated into Chinese. Taking it all in all, it seems probable that Lamarckian evolution, or the inheritance of any sort of acquired character, has played at most a minor and insignificant part in the actual evolution of animals and plants. That this is so is lucky for humanity. For, it is undoubted, and even obvious, that most human beings fall far short of reaching the standard either of body or of mind to which their inherent constitution could develop if it was made the best of. Either through their own fault, or through the fault of a bad environment, in the shape of poverty, slums, monotonous life, unhealthy work, or unhappy family circumstances, they are in some way or another under-developed.

If this under-development were inheritable, the majority of the race would be undergoing a steady degeneration. As it is, however, a good stock in a bad environment continues to produce potentially normal and healthy children, that only want good surroundings to unfold their possibilities.

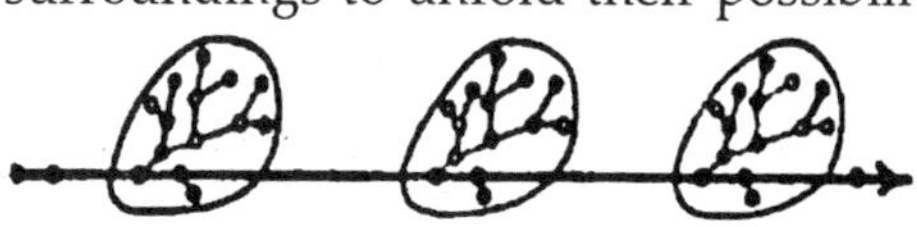

Fig. The Relation Between Body (*Soma*) and Germ-plasm in Successive Generations

Only the germ-plasm, represented by a succession of dividing germ-cells, is continuous from one generation to the next (black line with arrow). Single individuals are represented by ovals, containing somatic cells (white circles), all of which die, and germ-cells (black circles), some of which produce the next generation. The son is therefore not descended from his father's body (*soma*), but both father's *soma* and son are descended from a common ancestor in the shape of the fertilized egg which grew into the father.

The second great theory of the method of evolution is that of Natural Selection. Darwin himself saw it forced upon him as the necessary consequence of the following facts of biology:

Fig. Nine Varieties of Domestic Pigeon, to Show the remarkable variations of shape, proportions, colour, number of feathers in the tail, which have been produced by selection from the single original stock of the wild Rock Dove

1. The fact of heredity all organisms tend to resemble their parents.
2. The fact of variation two organisms are exactly alike. Thus the resemblance between parent and offspring is not absolute. Further, some variations at least are inheritable.
3. All organisms produce more offspring than can survive. If all, or even half the young, even of the slowest-breeding known animal, the elephant, were to come to maturity and themselves reproduce, the

whole globe would in a limited time become packed with elephants. While what would happen if all the million or two eggs of every sea urchin came to maturity baffles imagination. Thus, among the individuals of every species, there is of necessity a *struggle for existence*-not necessarily a conscious struggle, but none the less a real competition in effect.

The conclusion to be drawn is this that those individuals which possess variations helping them in the struggle will on the whole survive, while those which have varied in the opposite direction will on the whole be killed off. Those that survive will reproduce the race, and by the operation of heredity, their offspring will tend to resemble them in other words, will tend to possess the same favourable variations. This process Darwin called *Natural Selection.*

Natural Selection may thus be compared to a sifting of the individuals of a race, a sifting which results in the race coming to consist only of such individuals as are best *adapted to their environment*. Darwin clinched his argument by pointing to the effects of artificial selection. Look, he said in effect, at the different breed of dogs, of pigeons, of horses, the different kinds of vegetables, corn, or fruit.

A Shetland pony, a race-horse, and a cart-horse, exhibit much greater points of difference than do many natural species. If a zoologist who by some accident had never seen a dog were shown a greyhound, a pug, a mastiff, and a dachshund, he would probably classify them in different genera.

Fig. 'Mutual Courtship'

Yet they, as well as the types of horses and all the other breeds of domestic animals, have been produced by selection, the breeder keeping such animals as came nearest his ideal, rejecting the others and not allowing them to reproduce. If artificial selection can produce such results in the short period covered by history of man, Natural Selection, argued Darwin, in the vast periods of geological time could produce all the diversities seen in the animal and plant kingdoms.

Another form of selection, which as Darwin saw, must operate among the higher forms of animals, is that which is called sexual selection. Just as natural selection is the immediate outcome of the struggle for existence, so sexual selection is the result of the struggle for reproduction. Where, as in all higher animals, reproduction is sexual, any advantage in finding a mate will give an evolutionary advantage, since the more successful animal will leave more descendants, who will on the average inherit the same qualities which ensured their parent's success.

Thus, while in Natural Selection, the sifting mechanism consists almost exclusively of the inorganic environment together with the actions of the animal's organic competitors, in sexual selection it consists largely of the mind of the opposite sex of the same species. In other cases, however, the males fight for the females, and here strength, skill, vigour, and protective and offensive weapons will be the decisive factors. Characters originating in this way by the antlers of deer, the canine teeth of horses, the spurs of male game-birds, the mane of the lion, or the enormous size of the males in many seals.

However, the more interesting qualities developed by sexual selection are those which have had to go through the sieve of the mate's mind. As we should expect, the characters most encouraged by this means will be those which most readily please, stimulate, or excite the mind, especially on its emotional side. And, as a matter of fact, they are just such characters: brilliant colours like those of male pheasants, often combined with striking patterns and structures, as in the train of the Peacock; wonderful ceremonies or dances such as that of the Grebe or the Argus Pheasant, in which bright or beautiful plumes are shown off in a startling or unusual way; or again, pleasant scents, as those sprayed over their mates by some male butterflies, like the 'Blues'; or gifts of food, as are made by some spiders and insects; or the instinct to construct regular 'art galleries' such as are made by the Bower-birds of Australia; or love-songs, like those of grasshoppers or song-birds; or even special vibrations, like those indulged in by numerous males among the web-spinning spiders (a very necessary precaution, since if they did not vibrate the female's web in a special way, different from that produced by the struggles of captured prey, the female would in all probability kill her suitor before realizing her mistake)

Fig. The Cock Argus Pheasant in Ordinary Attitude

One thing is interesting. On the whole, the characters that have been evolved through sexual selection are either pleasing or interesting to us, showing that in general whereever there is a mind at work, it is pleased and stimulated by the same kinds of things that please and stimulate our minds. Often, it is true, the details of the display-character seem grotesque or even unpleasant to us, but this is generally due to some incongruousness.

Characters like these are usually developed only in the males, the selection being exercised by the mind of the females. In such cases, selection will be most effective where polygamy prevails, as in Black-game, since there the favoured male may leave a very large number of young, the unsuccessful males leaving none at all. In other cases, however, notably in birds in which both sexes share the duties of sitting on the eggs and looking after the young, and there is thus a real family life, the stimulation and the selection is mutual, and adornments are evolved in both sexes.

In such cases, the ceremonies and dances connected with display appear to have taken on an additional function, namely, of providing an 'emotional bond' which helps to keep the family together for the good of the species.

In any case, the evolution of sex, the final ousting of all other forms of reproduction by sexual reproduction, followed by the development of brain and mind to a high level at which emotions count in determining behaviour, made the operation of sexual selection inevitable; and this in its turn determined the evolution of a very great deal of the beauty and interest of the higher forms of life.

Lamarckism in the broad sense implies a direct action of the environment upon the constitution of the race. The Selection Theory, it will be seen, only demands an indirect influence of the environment, but in both cases the environment plays a great part in guiding the direction of evolution.

A third theory of evolution has been advanced, called Orthogenesis, or development in straight lines. In its strict sense, this view implies that evolution proceeds in any particular direction, not because of any advantage gained by the race, nor because of any direct moulding effect of the surroundings, but because of some inner urge, some necessity for the hereditary constitution to change in just that particular way.

However, many of the points adduced in favour of the theory turn out not to support one theory of the method of evolution more than another, but to be merely statements of fact—of the very interesting fact that evolution *does* often (as in the horses, for instance) pursue a slow and straight-line course.

In these cases, the theory of Natural Selection will serve equally well to account for the facts, without our having recourse to any new principle.

On the other hand, some very puzzling cases are known. For instance, the great group of ammonites, a division of molluscs related to the cuttlefish and nautilus, and now all extinct, were very abundant in the secondary period. Many of them, towards the close of their hey-day, not only evolved along straight lines, but along lines which seem to be dooming the evolving race to eventual extinction. It is difficult to account for this in terms either of Lamarckism or of Natural Selection.

It is clear that the range of possible variations is limited by the particular hereditary constitution: no insect can, it appears, ever develop bone, no vertebrate can produce a compound eye. This implies a limited form of Orthogenesis; and perhaps in certain cases the restrictions imposed by the nature of the constitution are greater. In any case, Orthogenesis can never be more than a subsidiary method of evolution.

Since Darwin's time, knowledge has increased in many ways, and has compelled us to revise or re-state some of his conclusions. For instance, we now know much more about heredity, about the way in which organisms tend to resemble their parents. We also know much more about variation. Some variations turn out not to be inheritable at all.

We have already considered some under the head of acquired characters; and, as a matter of fact, a great many, perhaps the majority of the minute differences which distinguish individuals are of this sort. In general, these non-heritable variations are called *modifications*. Opposed to them are variations which can be inherited. These are called *mutations*.

They may be very striking, as in the case of the Ancon breed of sheep mentioned by Darwin, where a single unusual ram suddenly appeared in a strain of ordinary sheep. This ram had short crooked legs and a long back like a turnspit dog. It transmitted its characters to its offspring, and from it a whole new breed of sheep was produced, the value of the breed being that owing to its inherited peculiarities it could not jump over fences.

Or mutations may be so small that they are masked by the effect of modifications and are only to be detected by special experiments. A good example is given by the weights of bean-seeds, a problem worked out by the Danish botanist Johannsen.

If a number of bean-plants are grown and their seeds weighed, the weights will range widely—from about 200 to

900 milligrams in this particular experiment. The seeds of any single plant, however, will have a much narrower range of weight. Further, the ranges of weight for the seeds of different plants may overlap, e. g. those of one ranging from 200 to 650, of another from 400 to 900 milligrams. If the seeds of one particular plant are saved, and grown separately, they will produce plants each of which will have the same range of seed-weight as did its parent.

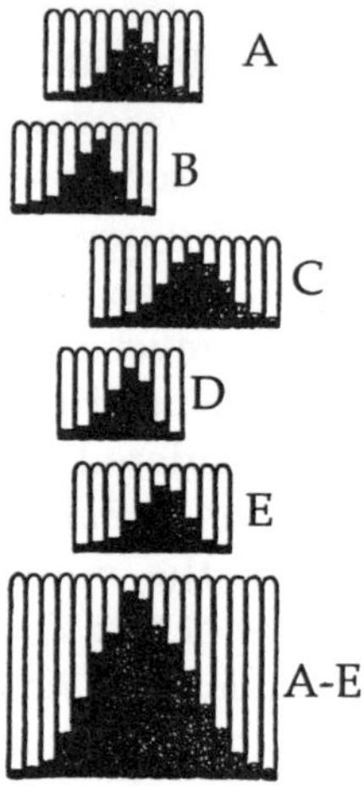

Fig. *A* to *E*, Five Pure Lines of Beans

This range of weight will be the same whether the bean which gave rise to the plant was light, medium, or heavy. Conversely, how-ever, two beans of the same weight, which have come from two plants with different ranges of seed-weight, will produce plants giving different ranges of seed-weight in their turn.

It is clear that there are in existence a number of hereditary variations or mutations affecting seed-weight. Plants possessing one of these will have seeds not of a definite and constant weight, but varying again (this time by modification) between fixed limits. The amount of the modification, however, does not affect the hereditary constitution, and the range of weight remains constant from generation to generation.

The average weight for each strain (under given conditions) is fixed by.heredity; but the differences in weight within the beans of one strain are due to environment—to the plant being in better or poorer soils, to the bean being at the base or the tip of the pod, the pod at the base or the tip of an upper or a lower branch, and so on. If selection for weight of seed were to be made among a large mixed lot of beans, and only those plants which had beans above a certain weight were used for breeding, we should get a rapid increase of average seed-weight in the population.

This would be because all the strains with low average weight, not being allowed to reproduce, would be eliminated altogether. After a few generations, however, selection would no longer produce any effect upon the average weight.

The reason of this would be that we were now dealing only with the strain with highest average weight, all the rest having been selected out; and each single strain breeds true to its average whether we select its heaviest or its lightest beans from which to breed. Further progress could only be made if a new mutation were to crop up (as might quite possibly happen, and as did happen during one of Johannsen's experiments) in the direction of greater seed-weight.

Recent breeding experiments have shown that new mutations are continually cropping up in all sorts of animals and plants, both wild and domesticated. For instance, in a species of the little fruit-fly Drosophila, the American zoologist Morgan has discovered over 400 new mutations, all of which have started from pedigree stock in the laboratory and all of which are inheritable.

Some of these involve very small changes only, such as a slight difference in eye-colour, while others, such as a mutation which makes the flies almost wingless and incapable of flight, are far reaching in their effects. Any one mutation may, it appears, influence any structure or function of the body in any direction.

Darwin's statement concerning variation must therefore be put in a new form. It should run thus.—All animals vary continually and in every part. A great deal of this variation is due to modification resulting from differences in environment, and is not inherited. Inheritable mutations, however, also occur which affect every part. In any given species a great number of these mutations are already in existence, and new ones are regularly arising.

Natural Selection, therefore, is a selection of mutations, large or small. At present we know very little as to *why* or how mutations originate. That is one of the greatest tasks before biology in the immediate future. But the fact that they *do* originate is sufficient to make evolution by Natural Selection possible and intelligible.

In this way, by the constant killing-off of more organisms with disadvantageous mutations, and the survival of more organisms with advantageous mutations, change will occur in the direction of better adaptation.

It must never be forgotten that a very slight advantage will in the course of generations come to preponderate. If for example, at the beginning of a glacial epoch, when the.climate were growing slowly colder, a mutation occurred in a particular kind of beetle such that it gave a 1 per cent. advantage to its possessor—in other words that on the average 100 individuals with the new character survived, but only 99 without it; suppose further that the new mutation and the old condition were simple Mendelian allelomorphs; and that the new mutations were dominant; suppose further that the mutation existed in 1 per cent. of the population. Then it can be calculated that after less than 500 generations, half the population will show the new character, and that in 1658 generations (not a long period in evolutionary history) 99 per cent.

That selection does take place in nature is known by many observations. To take a simple instance, after a severe storm, an American naturalist collected a number of sparrows which were lying exhausted on the ground. Some of these recovered, the rest died.

The wings of those that recovered were nearer the average for the species; the wings of those that died were on the average either much larger or much smaller than the normal. The theory of Natural Selection, in the slightly altered modern form that we have stated, is becoming more and more firmly supported by evidence as time goes on, and there can be now little doubt that it has been the most important agency in bringing about adaptive and progressive evolutionary changes.

Chapter 10

Processes of Evolution

One of the most important properties of animals and plants is adaptation to their conditions of existence. Of this obvious but most fundamental point, it will be as well to give a few examples from widely different groups of animals. Birds and mammals both possess a constant temperature which is ordinarily well above that of their surroundings. They should, therefore, be equipped with an arrangement for preventing too rapid loss of heat; and this is provided by the feathers in one case, the hairs in the other.

Swimming birds, almost without exception, are webfooted or lobe-footed. The Jacana, which seeks its food on the floating leaves of water-plants, has enormous toes which distribute its weight as do skis or snow-shoes on the thin crust of snow. Indeed, in general the feet of birds are remarkably well adapted to the life which their owners lead. The egg-eating snake, Dasypeltis, possesses a special mechanism by which it can temporarily dislocate its jaws to swallow an egg whole.

In addition, one of its neck vertebrae bears a downwardly directed spine which protrudes into the gullet, and is used to crack the egg where none of its contents will be wasted. The larvae of crabs live near the surface of the sea, and are provided with long spines which increase friction and make them less ready to sink. But the most striking examples are those known as convergence, where a similar mode of life produces similar effects on quite unrelated animals.

For instance, it is advantageous for many animals to be invisible against their surroundings, either to escape their enemies or to approach their prey unobserved. And we find that in polar latitudes many animals are white, at any rate in winter, in deserts many are sandy, in undergrowth many are blotched and streaked so as to break up their outline, and harmonize with the tangle. Others, again, escape their enemies by mimicry, as it is called by resembling other animals which are dangerous or distasteful. Thus wasps, which advertise their sting by their bright black and yellow pattern, are imitated (of course quite unconsciously) by a large number of different sorts of perfectly harmless insects both as regards pattern and bodyform. Ants may be mimicked by many other insects.

Fig. Protective Resemblance. Four Specimens of The Crustacean

Huenia proteus on the sea-weed Halimeda. The resemblance is striking both in form and colour.

Another type of convergence concerns shape. Every one is familiar with the typical fish shape—the stream lines of the body, permitting of rapid motion in water. When other vertebrates have taken to the sea, they too have evolved into a similar shape, as is seen in the Ichthyosaurs, a group of reptiles, and the whales and porpoises, a group of mammals.

Fig. Mimicry and Protective Resemblance in the East African Grasshopper

This adaptation of every part of an organism to its role, of every whole organism to its mode of life, is universal. It is indeed the direct and most obvious outcome of Natural Selection.

Fig. A Lantern-bug

But there is another fact, of perhaps greater importance, which must be taken into account, and that is the existence of higher and lower types of organisms. Within a few square yards we may have the man in a house, the dog in the yard, the worm in a flower-bed, the fish (probably a goldfish) in a pool in the garden, and the amoeba also in the pool, or in a water-butt.

They are all living close together, but their effective environments are amazingly different in extent. Most of the happenings to which the amoeba responds accurately take place within a radius of a millimetre or so; when stimuli like light or vibration affect it, it has no means of discovering anything about the distance or the kind of source from which they come, but responds simply to light or to vibration as such.

Fig. Views (Upper, From The Side; Lower, From Above) of Four Species of Plant

The total range of environment which the Hydra could conceive of would be a few centimeters each way. The worm is a little larger, but still without any special sense-organs; it can distinguish light from darkness, but cannot see the shape or colour or distance of anything; it can tell when the earth is vibrating, but cannot in any true sense *hear*, because it cannot distinguish tones; it cannot begin to control the environment in the same sort of way as man controls it, because it is not even in contact with most of that environment, locked away from the happenings of the outer world in a windowless existence which is almost incomprehensible to us who are provided with efficient sense-organs.

Fig. Side View of a Plant-bug (*Membracidae*) Mimicking an ant

The fish can see images, but its vision of anything outside the water is very limited owing to the water surface, which if it is rough prevents vision across it. In addition, it is of course confined to water, and so, in this case, to a little pool of a few feet radius.

The dog can see, can hear and smell very well, and can roam over the surface of the earth. Its environment as perceived by its senses is as extensive as that perceived by man's senses; but it cannot understand it in the same way. For instance, we can be pretty sure that no dog could ever come to understand the difference in distance between a cloud, the moon, the sun, and the stars. In addition, its brain cannot make the same reasoned relations between its sense-impressions as can man.

Although it can and does learn, it can only learn in an unintelligent way, making associations as they come. It does not appear capable of thinking in abstract terms of cause and effect. Its environment must seem both more limited and more chaotic than the man's. The man, if he takes the trouble to be interested in the environment in which he lives, finds it a very marvellous one.

He can get information, by means of a microscope, of invisible things under his nose; he can also obtain information, by spectroscope and telescope and mathematical calculation, about the composition, size, and speed of stars hundreds of light-years away. He can know by letter and newspaper what others are thinking and doing, through history he can enlarge his past far beyond the limits of his single lifetime, and he can make prophecies, sometimes (like those of eclipses and comets) of perfect accuracy, concerning the future.

Also he can relate the different parts of his environment to each other and to general principles. His environment is enormously greater, both in space and time, than the dog's—let alone the amoeba's; and it is much more intelligible.

In addition, the six organisms are of very different sizes and degrees of complication. Amoeba consists of one cell, hydra several thousand, the worm many hundreds of thousands, a man millions of millions. There are many more kinds of cells in man, dog, or fish, than in hydra or even worm. As we ascend the series, we find even greater independence of external forces.

Amoeba and hydra are at the mercy of floods, currents, droughts. A worm is limited to a very small section of the soil. A fish is wide-ranging, and can change its abode at the onset of unfavourable circumstances, but is confined to water. The dog can range on land, and, finally, man is at home in every latitude, and has mastered sea and air as well as earth.

When we look carefully into the matter too the differences between members of the series can be thought of in relation to the environment, but in a much more general way than is the case with special adaptations. Using the word environment to denote the whole series of events and processes with which life can come into contact, and not merely the particular environment of one particular organism, we can say that some animals have more control over the environment than others, and are more independent of it.

The savage is more at the mercy of the elements than is civilized man; civilization is learning to control floods and droughts, to make a pathway of communication of the same sea which to the savage is an impassable barrier. The dog in its turn is not able to cultivate the ground or to kill such varied game as the savage. The fish can exert far fewer different movements than the dog, and is much less able to profit by experience. The earth-worm is not only without specialized organs of locomotion, but also lacks all organs of special sense. So far as independence goes, it is well to remember that amoeba and hydra are bathed, over the whole of their absorptive surface, by the water in which they live, and that accordingly any changes in the composition of that water act immediately upon the vital processes of the animals. In addition, they do not possess any mechanism for regulating their temperature, and so must live slow or fast according as their surroundings are cold or hot.

In man or any mammal or bird, both the temperature and the chemical composition of the fluid which is in contact with all the cells of the body are kept constant with an extraordinary degree of accuracy, and special devices exist for preventing changes in the outer world from exerting their full effect upon the vital processes of the body.

In brief, we may say that high and low organisms can be distinguished by the degree of their control over and their independence of environment. This difference in independence and control is reflected in their structure and their capacity for self-regulation and, in the mental sphere, by the degree of knowledge of the outer world which their sense-organs and brains permit, and in all probability by the intensity of their emotions.

From what we know about evolution it is clear that the highest organisms have developed latest in time. This we can actually see happening as we trace back the history of life in the fossils; it is a probability which amounts for all practical purposes to certainty that the converse is true, and that although the early stages of the earth's history as written in the rocks and fossils are now undecipherable, yet that the first forms in which life appeared were low organisms.

The total thickness since the beginning of the Cambrian is thus about 36 miles. Theincisions indicate the periods when various great mountain-systems were formed—Eurasiatic on the left, American on the right. The dominant forms of animal life areindicated on the right; it will be noted that highly complex forms (Trilobites) had already been evolved in the Cambrian.

At the beginning, then, there were only low organisms, today there exist all gradations between the highest organisms and the lowest undoubtedly many organisms have degenerated during evolution from a higher to a lower condition. If we look back into the history of fossils, and investigate what forms of life were present before and after the development of some new high type, such as man or the mammals, we shall almost always find that the new type has simply been added to the previously existing types.

For instance, the reptiles were the dominant land animals in the secondary period, before the development of the higher or placental mammals; when these were evolved, just before the beginning of the tertiary period, although they speedily became dominant on land, and although many species and even whole sub-groups of reptiles were extinguished, yet the reptilian type as a whole did not perish from off the face of the earth, but continued to exist as well as the mammals.

In the same way, although the advent of man sealed the death-warrant of a great many species of other mammals, reduced the total number of lower mammals very considerably, and deposed them from their previous dominant position, yet lower mammals still exist in abundance, and will undoubtedly continue to do so.

Thus we cannot say that evolution consists simply in the development of higher from lower forms of life; it consists in raising the upper level of organization reached by living matter, while still permitting the lower types of organization to survive. This general direction to be found in evolution, this gradual rise in the upper level of control and independence to be observed in living things with the passage of time, may be called *evolutionary* or *biological progress*. It is obviously of great importance, and can be seen, on reflection, to be another necessary consequence of the struggle for existence.

This improvement has been brought about in two main ways, which we may call *aggregation* and *individuation*.

Individuation is the improvement of the separate unit for example, in the series Hydra—Earth-worm—Frog—Man. Aggregation is the joining together of a number of separate units to form a super-unit, as when coral polyps unite to form a colony. This is often followed by division of labour among the various units, which of course is the beginning of individuation for the super-unit, the turning of a mere aggregate into an individual.

Let us take as an obvious example of biological progress the colonization of the land by vertebrates. As a matter of verifiable fact, the sea was already peopled with highly developed fish before the first amphibians appeared on land. In other words, while there existed great competition among vertebrates in the sea, this competition did not as yet exist on the land. Clearly, then, it would be a biological advantage to any species if it were to vary in such a way as to make it able to live on land, for then unchecked multiplication would be possible for it, and it would have fewer enemies. Variations in this direction would thus tend to be preserved; in other words, this particular step in biological progress would be favoured.

As a matter of fact the step is a very large one, and progress was inevitably slow. The Amphibia did not arrive at a complete solution of the problem of leading a terrestrial existence. Their skin is moist, they are usually confined to wet places even in their adult existence, and the earlier part of their life is almost invariably spent actually in water, in the shape of a tadpole larva.

With the evolution of the Amphibia the fringe of the dry land, the territory between land and water, had been conquered, but not the dry land as a whole. Once again, after millions of years during which the Amphibia were the highest vertebrate type, evolving life was confronted with a situation in which a premium was placed upon a further advance: animals born with heritable variations making it possible for them to live farther and more permanently away from water would become heirs of a rich unoccupied territory. Thus it was that the Amphibia, after themselves arising from the fish, in their turn gave origin to the reptiles. In the same way, in the continual struggle that is going on in mammals or birds between herbivore and carnivore, pursuer and pursued, each new advance in speed and size or strength in one party to the conflict, must call forth a corresponding advance in the other, if it is not to go under in the struggle and become extinct. The striking improvement in the running powers of the horse family during its evolution, evinced in increase of size, lengthening of the legs, reduction in the number of the digits, and development of a well-formed hoof, and the similar improvements that occurred in other Ungulates, were accompanied by a corresponding increase in the size, speed, and power of the group of Carnivora during the same geological periods. Each was at the same time the cause and the effect of the other.

A precisely similar state of affairs is often to be seen in the evolution of the tools and weapons and machines of man. For instance, in naval history, the increase throughout the nineteenth century of the range and piercing power of projectiles on the one hand, of the thickness and resistance of armour-plate on the other, provides a very exact parallel with the simultaneous increase of speed and strength in both carnivores and their prey.

In Nelson's time, the men-of-war were built of unsheathed wood, and the guns fired round iron balls, with a maximum range of a few hundred yards. To-day, battleships carry 15-inch guns which fling steelcapped and pointed projectiles, laden with high explosive, for a dozen miles or so, while the armour-plating of heavily protected ships may now reach a thickness of a foot or even more of specially-treated nickel steel.

The earliest types are placed below. The number of digits on both limbs is slowly reduced; the middle digit is enlarged and specialized as a hoof; the fore-arm is strengthened by the fusion of radius and ulna, the ulna in its distal part disappearing; the hind limb is strengthened by the disappearance of the fibula and corresponding enlargement of the tibia; the length of the tooth is increased; its grinding surface becomes more complex.

In addition there is an increase of size and change of proportions. During the interval, progress has been steady and gradual in both departments of naval warfare, each advance in efficiency of guns being the stimulus for new invention in the methods of protection, and vice versa.

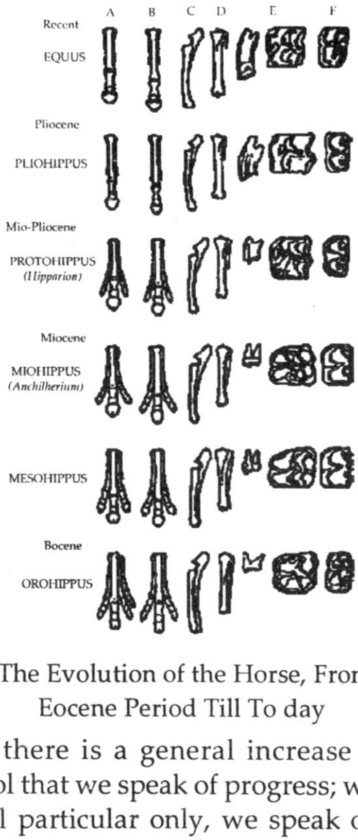

Fig. The Evolution of the Horse, From the Eocene Period Till To day

It is when there is a general increase of the animal's powers of control that we speak of progress; when the increase is in one special particular only, we speak of specialization. For instance, the horse is specialized for running, the mole for burrowing, the bat for flying, the whale for a marine life, the lion for catching and devouring large animals, the sloth for living in trees.

Each of these animals is well adapted for its particular mode of life, but each is by that very adaptation quite cut off from leading the life of any other. During the late Secondary period there existed similarly specialized types of animals. For example, the Ostrich Dinosaur was adapted for running, the Ichthyosaur for life in the sea, the Tyrannosaur for preying on large animals, the Pterodactyl for flight, and so on.

But all the former list of animals are mammals, all the latter were reptiles. The mammals are higher in that they possess proper temperature-regulation, for instance, and better prenatal care of their young, as well as in many other points. Thus, while the types of specialization, or of adaptation to particular modes of life, are somewhat similar in the two cases, yet each member of the first group is higher than any member of the second, because its general organization is on a more efficient level.

Whenever a new group of animals is evolved, it is found that its members soon become specialized in different directions, thus filling up different vacant places in the economy of Nature. This adaptation to different modes of life, while we call it *specialization* when we are thinking only of one species of animal, is called *adaptive radiation* when we are thinking of the group as a whole. All fish, for instance, breathe by gills, possess fins as limbs, and have other characteristics in common, so that any member of the group can be at once recognized by a brief examination of its structure.

Yet the detailed form of different species of fish is extremely varied. Besides the ordinary type of active, free-swimming fish, like the herring or trout or mackerel, there are fish flattened in adaptation to life on the bottom, some flattened sideways, like the sole and plaice, others flattened from above downwards, like the skates and rays; there are elongated fish like eels and pipe-fish; there are fish with prehensile tails, like the sea-horse; there are the flying-fish adapted for leaping long distances out of the water; mottled fish of irregular outline adapted for living on rocks; deep-sea fish with wonderful phosphorescent organs for searchlights and huge eyes for perceiving the faintest trace of light; cave-fish without eyes at all; sucker-fish adapted for sticking tight to the underside of stones, or for being carried about by larger fish without expending any energy themselves; and so on and so forth through almost every conceivable sort of form possible in an under-water existence. The examples previously mentioned give an idea of some of the adaptive radiation which has taken place in reptiles and in mammals; very similar instances could be taken from any other large group, such, for instance, as the insects. It is interesting to take any such group and to see what part adaptive radiation has played in giving rise to the main sub-groups into which it is divided.

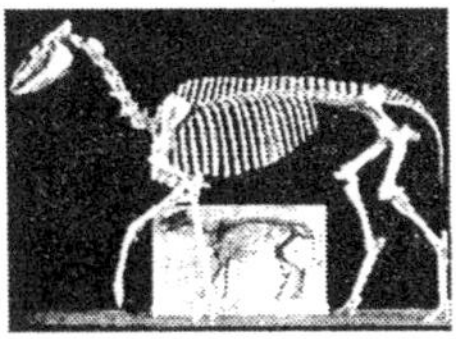

Fig. Photograph of the Skeletons of the Small Ancestral Horse

There is one particular form of specialization which we have not so far mentioned; and that is the retrograde form of specialization known as degeneration. There are many cases known where animals can be definitely shown to be less highly organized to-day than were their ancestors in the past.

The animal known as Sacculina, for instance, is a parasite upon various sorts of crabs. It consists of a mere bag filled with little else but reproductive cells, and sending out a whole series of branched roots which penetrate the crab's body in all directions and suck nourishment out of it. At first sight, the relationships of this unpleasant creature are very difficult to determine. But when its development is investigated, it is found that it hatches out of the egg as a a free-swimming larva exactly like those found in many Crustacea.

It has jointed limbs, an external skeleton made of chitin, and in fact bears all the distinguishing marks that other Crustacea do, be they crabs or lobsters or shrimps or water-fleas. It is, in fact, a crustacean which has become adapted to a parasitic life; and in so doing it has acquired special adaptations, such as the root-like organs, specially suited to that life, while it has lost sense-organs, limbs, digestive system, and everything else necessary for leading a free and independent existence.

It has lost more than it has gained, its organization has

become simpler, its independence less; in fact it has gone down the evolutionary hill, and the direction of its history has been in most ways the opposite of the direction which characterizes biological progress.

The form in which Sacculina hatches out resembles in general many crustacean larvae; but it is particularly like the early stages of the animals known as barnacles. Every one who has been to the seaside knows what an acorn barnacle looks like—a little creature attached to rocks or piles, enclosed in a white shell, and capable of sending out of a slit in the top of the shell a regular sweep-net composed of a number of 'arms'—really appendages—with which it drags minute floating particles of food in towards its mouth.

Even in the adult state a barnacle shows some resemblance to ordinary Crustacea, especially in its jointed limbs and chitinous external skeleton; the early free-swimming larva clinches the matter and gives complete proof, as in the case of Sacculina, of their crustacean nature and affinities. The very close resemblance of the larva of barnacles to that of Sacculina points to an especially close connexion between the two sorts of animals; and as a matter of fact, the two are undoubtedly descended from a common stock.

The barnacle is degenerate as well as the Sacculina, but it is not so degenerate. It still, for instance, possesses organs for capturing and digesting food. On the other hand, it has lost its organs of special sense and of locomotion. Further, it is not adapted to the same general mode of life as is Sacculina; it is not parasitic, but sedentary or sessile.

This settling down and becoming fixed is the other great cause, besides parasitism, of degeneration in animals; as would be expected, the degeneration due to a sedentary life is rarely so great as that due to parasitism, since the sedentary animal does not obtain its food ready-digested as do most parasites.

It must not be supposed that because the general rule among animals is that time brings change, that therefore time *invariably* brings change. The common lamp-shell Lingula, for instance, has persisted without any appreciable change whatever in the structure of its shell for the prodigious period of time, certainly over five hundred million years, which has elapsed since the Cambrian epoch in the Primary period. Even when individual species have changed, the general characters of groups have often persisted with very little modification, as, for instance, those of dragonflies since the coal-measures, of shark-like fish since the Silurian.

In some cases this may mean that for some unknown reason the species or the group has lost the power of varying to any considerable extent. More often, probably, it so admirably fills one particular niche in the order of things, and a niche which stays the same throughout the geological periods, that it pays for the animal or the group of animals to stay as they are, leaving it to other groups or to other branches of the same group to colonize new niches and to progress towards fuller existence.

To sum up, we may say that two main types of evolutionary change result in animals (and also, as a matter of fact, in plants) from the struggle for existence and the constant appearance of inheritable variations. In the first place, once a new type or plan of structure has been evolved, it undergoes adaptive radiation; in other words, there are developed a number of separate species all built on the same general ground-plan, but adapted to different and usually incompatible modes of life. In the second place, new types and new plans are continually appearing as time goes on, and progress is marked by the fact that among the later-evolved types there is to be found greater complexity of organization, greater control and independence of environment, than among the earlier.

It might perhaps be thought that specialization was often the same thing as progress. Specialization, however, implies close adaptation to one particular mode of life, while progress means greater general efficiency. If we look into the actual history of animals in the past, we find that specialization in one direction involves the sacrifice of possible advance in other directions, and is a barrier in the long run to any but a quite limited degree of progress.

As a result of its long course of specialization, extending

for tens of millions of years through the better part of the Tertiary epoch, a specialization all tending towards greater efficiency in running and browsing, the horse stock has, it seems, cut itself off from the possibility of adapting itself to other modes of life—to a life in the water or in the trees, or to a carnivorous habit.

There is a limit to the perfection which any line of specialization can attain. While the horse was growing larger, developing hoofs, reducing the number of its digits, another and wholly different type was being evolved in the person of man. If it were not that horses are useful to man, and are accordingly domesticated, they would now be wholly or almost wholly extinct.

The 'natural' enemies of the horse are the large carnivores. These are built on the same general plan as the horse—the mammalian—and indeed are the results of the specialization of the same plan in another direction. The same limits are thus set to them as to the horse stock. Before the advent of man, a state of equilibrium existed between the horse and its enemies, the latter not able to destroy the former entirely, the first not able to escape the payment of some toll to the second.

But the horse came up against the wholly new biological conditions introduced by the new type, man; it was in direct conflict with man's cunning and tools and his habit of hunting in bands; still more important, it had to compete with the indirect effects of the new development, such as the settlement and cultivation and enclosure of the land. Against all this the horse was powerless. It could not develop far enough or fast enough to adapt itself to such sudden changes, and as a result it is becoming extinct as a wild species.

Over and over again the same thing happens, and the specialized representative of the old type, plant or animal, is extinguished in competition with the specialized representative of the new. For tree-like representatives of the horsetail type, which existed in the Carboniferous period, we have seed-bearing trees to-day; for pterodactyls, birds and bats; for dinosaurs, the large mammals; for the large early amphibians, the Stegocephalia, we have crocodiles; for the abundance, both in numbers and in species, of the larger mammals in the Pliocene, we have the multifarious activities of the swarms of man.

The new type seems always to have arisen from some comparatively unspecialized branch of the old, and to have attained its pre-eminence by means of adaptations towards general instead of towards special efficiency. Man has ousted the other mammals from their dominant position owing to the development of his mind. Through his particular type of mind he is able to deal rationally with the problems that confront him; tools, machines, tradition, civilization, and unexampled control and independence have been the result. The human mind is not merely adapted for solving one or two particular problems; it represents a *method* more efficient than any previously adopted for dealing with any and every problem that may confront an organism.

Man's body is not highly specialized, and he seems to have arisen, through a monkey-like ancestor, from some unspecialized early mammalian stock like the Insectivores. Nor did the early mammals show any signs of specialization. All the fossil mammals that we know of during the time when the great period of reptilian dominance and specialization lasted were small, primitive creatures, at first sight not likely to wrest the palm from their powerful rivals.

Very similar chains of events may be seen taking place in the evolution of human machines and inventions. Take, for example, the history of transport. The most primitive method was the carrying of single loads by human beings or pack-animals. After that came the invention of vehicles in general and of wheeled vehicles in particular. The wheeled vehicle became specialized ('adaptive radiation') in innumerable ways.

We have the war-chariot; the rapid vehicle for passenger transport and for pleasure; the heavy wagon and cart; the van and pantechnicon; the four-inhand mail-coaches bowling daily at fixed hours along the main roads of the land. A limit was set to the capacity and the speed of such vehicles by the speed and strength of beasts of burden on the one hand, and by the imperfections of road-surface on the other.

At the beginning of last century a new type of vehicle was evolved for which these limitations no longer existed. It was discovered how to replace the energy of animals by that of steam, and in large part to overcome the difficulties of surface friction by making the wheels of the vehicles run on metal rails. As a result, steam locomotives became for certain purposes the 'dominant type' of vehicle within an extremely short period of time.

Here again the type itself has been improved, so that we have now for some time been close to the limit of its possibilities. It does not seem possible to run a profitable service at speeds of much over sixty miles an hour. About half a century later a new plan was evolved. The internal combustion engine was produced, and gave certain great advantages, notably in being ready to start at once without the long preparation of 'getting steam up'. It appears to be the fate of new types to lead a precarious existence for a considerable time before they can compete successfully with the dominant types of the period.

This was so, as we saw, with the earliest mammals; it was so for the steam-engine; and it was so, to a very marked degree, for the internal combustion engines. They were laughed at when their inventors took them out on the roads; the law laid down that they should be preceded by a man with a red flag; the early defects in construction did actually occasion many a breakdown. But within thirty years they came into their own.

Meanwhile, still another competitor is in the field—the flying machine, a totally new type, abandoning not only the particular device of the wheel, but the whole element to which the wheel was adapted. It looks as if in certain respects, where speed is the main object, the aeroplane would become the dominant type of vehicle; but that it would leave the major part of transport to be dealt with by train and by motor.

Another interesting parallel with the evolution of organisms is found in the fact that although there has been progress, although the dominant type of vehicle has altered with the passage of time, yet many representatives of the old types have survived. They have managed to survive by becoming restricted to a few special conditions and places. Packanimals, for instance, while once universal, are now only employed in mountainous or roadless countries. The American buggy is still of the greatest use over the unmade roads which are still to be found in so many parts of the United States.

The horse-drawn vehicle will long hold its own in businesses in which not much capital is available, and speed and great power are not of the first importance.

This survival of all or almost all the types that arise in evolution, even though new types arise and supremacy changes hands, is of general occurrence in the development of animals and plants, and is at first sight very puzzling. However, the insight which is gained by looking, as we have done, into the evolution of something familiar and human like the means of locomotion, helps us to understand the more complex and slower-moving processes of organic evolution.

Another point which is brought out by the study of the development of some human contrivance such as the means of transport, is the great speed of change now possible in human affairs as against the slowness of change prevailing in lower organisms.

The whole period from the stage-coach to the aeroplane is comprised in well under two centuries. The resulting change in human habits has been enormous; to produce comparable changes in the habits of an animal stock there would be needed a period certainly to be reckoned in tens of thousands, possibly in millions of years.

On the other hand, the evolution of machines is a perfectly real evolution. Two different types of machines capable of performing the same general function—such, for instance, as the motor-lorry and the goods steam-engine—do come into a very real competition with each other, and the issue of the struggle is decided by a form of true natural selection, depending in the long run upon which of the two pays the better.

Here again the study of machines throws light upon the course of events in animals. It is often supposed that evolution must involve some conscious effort on the part of the evolving organism, that the struggle for existence is a conscious struggle, or that a species in some mysterious way 'learns' how to develop some new improvement in its structure.

As a matter of fact, almost the whole of such ideas are purely metaphorical, and arise simply because we read the processes of our own minds into the operations of nature; it is not scientifically correct to speak, for instance, of *purpose* except in relation to human minds. We see at once that the machines have no idea themselves of the direction of their evolution, that the 'struggle' between them is only a metaphorical struggle, that the selection between them is so far as they are concerned a mere sifting process, the issue of which depends upon the advantages or disadvantages which they may happen to possess.

It is the same with organisms. If two races of animals come into competition, the issue is decided by the qualities which each happens to possess; 'natural selection' is a name for the effect exerted by all the forces of the environment with which they come into relation, an effect which acts again like an automatic sieve, and lets some through to perpetuate themselves, keeps back and so extinguishes others.

The 'struggle' and the 'competition' are again usually metaphorical only. For instance, when the common housesparrow was introduced into America, it entered into competition with many of the small sparrow-like birds which had developed in that country. But the struggle did not take the form of a war between the invader and the original inhabitants.

It was an indirect struggle, due to the fact that both lived upon the same sort of food, both occupied the same sort of sites. The European sparrow happened to be endowed with qualities which gave it an advantage, and as a result it has spread enormously over the American continent, while many of the native birds have correspondingly decreased. If we wish to use a human metaphor, we can say that its success has been the result of 'peaceful penetration', not of fighting.

The difference between machines and organisms, of course, is that the machines are directly designed by man, whereas the place of the designer in animal evolution is (roughly) taken by the variation which seems to be universal in organisms. It is variation which provides the raw differences upon which the sieve of natural selection can work.

Mention must also be made of the theories of evolution which are summed up under the term *orthogenesis,* which means evolution in straight lines. It is frequently found, as for instance in the development of the horses or of the elephants, that evolution as revealed by fossils proceeds straight onward through geological time in a perfectly definite direction-in the horses towards single toes and hoofs, in the elephants towards great bulk, tusks, and trunk. Orthogenesis is sometimes used merely as a descriptive term, to denote this observed fact of straight-line evolution.

But by others it is used to mean that there exists some inner necessity for the evolutionary line in question to develop in just that one way and no other. However, a series of fossils, even if beautifully complete, can really give us no insight into the method by which its evolution occurred. Whenever the direction in which the series is evolving is adaptive or seems biologically advantageous, the orthogenetic series can be perfectly well explained by natural selection. On the other hand, there do exist cases where at least no advantage can be perceived by us in the direction pursued.

This is so, for instance, as regards the Ammonites, the extinct cephalopod molluscs which died out near the end of the Secondary period. Near the close of their time on earth, they often evolved orthogenetically into the most bizarre forms, their spiral shell becoming unwound, or irregular. Possibly in such cases a real causal orthogenesis, an inwardly determined mode of variation, is at work. But we are justified in saying that such cases, if they occur at all, are certainly rare.

Chapter 11

The Animal Kingdom

The thousands of different kinds of animals that exist, and are apt to become overwhelmed with the detail. Each is unique, full of interesting points; one type may be extremely different from another. It looks at first sight as if we were dealing with facts in masses too great to be properly assimilated.

But, while the facts are in a sense infinite and our understanding of them must always remain incomplete, yet we can reduce the apparent chaos to order, can get some mental grip on it as a whole, by means of a few general ideas and principles which lie ready to our hands.

For the present purpose the general principles are three, all depending upon the central idea of Evolution. First, if evolution has taken place, some animals will be closer bloodrelations, some more distant; and by examining likenesses and differences, we can obtain a measure of the relationship. By classifying animals according to their likenesses into groups of various sizes, we can put any particular species in a labelled pigeon-hole, so to speak, where it can always be found again if it is wanted. What is more, since such groups represent collections of *related* species, our method of pigeon-holing has reality behind it, and we have a Natural Classification.

Working on these lines, zoologists classify the whole animal kingdom into about a dozen main groups or *Phyla*. The most familiar phylum is that of the backboned animals, to which man belongs—the Vertebrates, or more accurately Chordates—while the insects belong to the group of animals with many-jointed limbs, the Arthropods; the jelly-fish to the group which possesses only one cavity to fulfil the function both of gut and body-cavity, the Coelenterates; and so forth. Each phylum is subdivided into classes, the classes into orders, and so down to families, genera, and species. In this scheme every animal has two names, that of the genus coming first, and serving as a surname, that of the species following, serving as a christian name.

This method of naming was invented by Linnaeus, the great Swedish naturalist of the eighteenth century. Before his time some animals were known only by the single familiar name like cat or pig, which led to confusion between different kinds of cats or pigs; or, in an attempt to remedy this, by a regular description (just like the long 'name' of the village in Anglesea which is really a description, and has been cut down by the practical post-office to a real name, Llanfair P.G.).

We must give names for things before we can know what we are talking about, and Linnaeus's system of naming animals and plants was a great step forward.

In this way, then, when any species has been properly looked at and investigated, it can be put in its right place in the scheme, and if we know certain essentials about the characters of the main groups, we know a good deal about the sort of animal which the particular species is, and also about its relatives, their number and chief characteristics.

The second (as also the third) principle which will help in ordering the facts, is concerned not with the animal's bloodrelationships, but with its interconnexions with its environment. Animals may evolve, but their evolution must always bear a close relation to the forces around it. The result of this is *adaptation*: animals (and of course plants too) always show some suitability to the particular conditions in which they live.

They also are fitted to the general conditions which characterize our particular planet. The planet Jupiter is so much larger than the Earth that the force of gravity on its surface is about 2·64 times as much as with us. If life could exist on its surface (which at present it probably could not) this difference in the force of gravity would have necessitated various special developments. Any Jovian land animals would have to be of a very different type from those on Earth; creatures like horses or deer would be physical impossibilities, and great stumpy legs would be a necessity for existence.

As to aerial winged animals, they might be wholly impossible; in any case, none using any flying mechanism found among the aerial forms we know could reach more than a very small size.

Or again, no known animal has any special sense-organ for detecting X-rays, or any protection from X-rays, although rays of this short wave-length are very harmful to life. For the simple reason that X-rays do not exist in nature on the Earth.

We have not even any sense-organs for telling a live rail or wire from one along which no electric current is passing, and many deaths occur each year in consequence. Although very big alterations of the environment are there all round the current, we cannot perceive them because in nature electric disturbances are either trivial in extent, or else occasional and uncontrollable like the lightning.

The third principle has already been touched on *biological progress*. It is in a sense a special case of adaptation, but it is of importance since it enables us to think of higher and lower animals as well as mere relationship or special adaptation.

We suppose then that life, starting in some very simple and probably tiny form, has gradually evolved since the day when it first appeared on earth, probably between one and three thousands of millions of years.

It has evolved into the huge number of species close to a million that are known to-day, as well as those, probably a far greater number, which have become extinct on the voyage through time.

The fundamental characteristics of life, its power of metabolism, assimilation and consequent growth and reproduction, are found in all creatures. Their other characters, almost without exception, are the outward and visible sign of the mode of life they need, imposed upon them by the necessities of existence.

With these main ideas in mind we may turn to the actual

Animal Kingdom as it exists to-day and in the record of fossil species, and see what we can make of it. Nothing certain is known probably nothing ever will be—about the form in which living matter originally existed on Earth. It is pretty clear, however, that the units must have been small, and that one of the earliest types of organization evolved was what we may call the cellular, in which the whole animal or plant consisted of a single cell with a single nucleus.

This fundamental type, although with great diversity of detail, is found in the great group of Protozoa, all of which are essentially single cells or quite simple colonies of cells.

Such a unit can never be much enlarged, for the simple reason that as its mass that which has to be fed and provided with oxygen increases as the cube of its radius, its surface through which its food and its oxygen must come increases only as the square. If a Protozoan should treble its diameter without changing its shape, it would have multiplied its surface by nine, but its bulk by twenty-seven, three times as much.

It is as if an island depending on imported food were to increase the population to be fed without a corresponding increase in docks, ports, and other import facilities. In spite of small size, however, some Protozoa attain very considerable complexity.

Various methods for circumventing this limitation are found among Protozoa. In some the body becomes elongated, in others partly divided into a series of chambers, in others, again, flattened and ribbon-like; the number of nuclei in the cell is often correspondingly increased. But this never leads to any size which we could dignify as even moderate; and few are the Protozoa which are visible to the naked eye.

A great number are parasitic, and some are the active causes of serious diseases, such as malaria, sleeping sickness, and one kind of dysentery. Some are even parasitic upon other Protozoa.

Reproduction is by fission, usually, as in the cells of our own body, by equal fission; sometimes by multiple fission. In most species at least, a form of sexual reproduction occurs at regular or irregular intervals. It is interesting that in a number of cases the conjugating cells, instead of being sharply distinguished into male and female, as with the sperm and eggs of higher forms, are nearly or even wholly alike. Thus, the Protozoa teach us that the essential thing about sexual reproduction is not the existence of two sexes but the union of two nuclei into one.

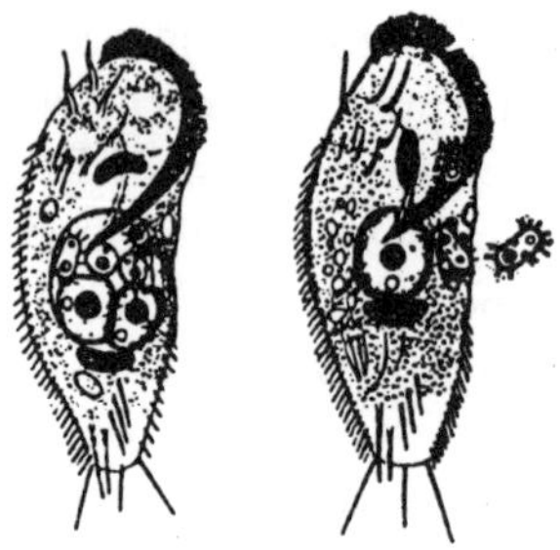

Fig. (*A*) The Complexity Which May Be Attained by Unicellular Organisms, (*B*) Parasitism in Protozoa

One of the facts about Protozoa most curious at first sight is that in them natural death does not exist, or at any rate only affects small parts of the body. When they divide, or when they undergo sexual fusion, nothing is left which can be compared to a corpse, and the substance of the original individual becomes directly transformed into two new individuals. The only death is accidental death, due to enemies or to bad conditions. Natural death has not yet appeared.

Increased size (up to a certain but very large limit) is obviously in many ways of biological advantage. A larger organism is less at the mercy of external agencies, less liable to the attacks of enemies. It is only when bulk is very great that serious disadvantages arise. Many bacteria are so small that they still show 'Brownian movement', in other words, their mass is so little larger than that of a molecule that the random movements of the molecules of the liquid in which they live can batter them and throw them from side to side. In most Protozoa and all higher animals this is no longer possible.

The force of surface-tension is enough to catch and hold fast many small insects if they fall on to still water, while no vertebrate has to face this inconvenience. Or again, the fastest-swimming Protozoa, or the little Crustacea called Copepods, or the larvae of crabs and lobsters, are powerless in the ocean currents which a herring or a mackerel breasts with ease, while the elephant or the deer crashes through saplings each of which is a world to hundreds of small creatures.

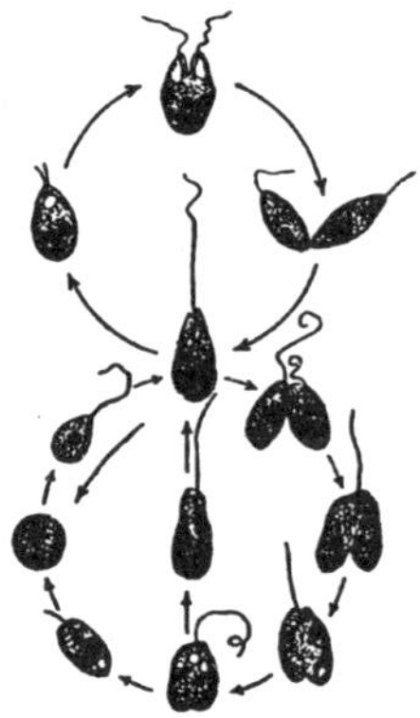

Fig. Structure and Life-cycle of a Protozoan

A method adopted by many Protozoa for reaping some of the advantages of size without the necessity for enlarging the individual cells is the formation of colonies. When fixed, these can produce strong feeding-currents by the united action of all their cells, and can raise themselves on a common stalk beyond the competition of the common herd of noncolonial unicellular animals and plants on the surface of the same stick or stone, while, when free-swimming, they can develop greater speed.

Fig. A Protozoan Colony, Without Division of Labour

Sometimes a division of labour is found between different

individuals of the colony, some serving for locomotion and feeding, others solely for reproduction. When this is so, it is really hard to say whether the colony is to be considered as a simple colony, a mere aggregate; or as itself individual-a compound individual of a different grade from the single cells which compose it.

Although the actual links are missing, there can be little doubt that all higher animals arose from Protozoa in this kind of way—by the aggregation of a number of cells into a colony, followed by a division of labour in the colony, first between reproductive cells and the others, which now merit the name of *somatic* or body-cells, and finally, by a further division of labour among the somatic cells into two layers, the outer protective and the inner nutritive.

Fig. A Colonial Protozoan, Zoothamnium

This particular step in evolution seems to have been taken twice over by different animals, leading on in one case to the sponges, in the other to the whole of the rest of the animal kingdom.

Sponges are almost unique among animals, for they have no mouth. They feed, as do so many of the smaller aquatic creatures, on microscopic particles of food extracted from a current which the animal passes through its body. In a sponge the current is sucked in through a great number of microscopic pores, and shot out at a single large opening, the *osculum*.

The cells that make the current are of a strange and interesting type, found nowhere else except in one small group of Protozoa. They each have a single actively beating flagellum, and are called collar-cells because this is surrounded by a delicate, living, transparent funnel or collar, which seems to help entangle food-particles.

Each of the collar-cells feeds separately from all the others; there is no proper digestive cavity, no common function of digestion for the whole animal. In this the sponges betray how little they have advanced from a mere colony of separate cell-units. The products of digestion diffuse out from the collar-cells to other parts of the sponge, or may be transported by special wandering cells.

No sponges possess any sense-organs or nerves, and the only movements they can execute are slow closures and openings of the osculum and pores. As might therefore be expected, they are all permanently fixed to the bottom throughout adult life.

Their characteristic form is maintained by a skeleton of spicules or fibres. Our bath-sponges, like many other species, are colonial, composed of a large number of sponge-bodies aggregated together. This can be seen from the fact that they possess many oscula instead of only one.

The sponges thus represent an early side-line in evolution, along which life never developed far. They are often put in a group Parazoa, as distinguished from the true Metazoa or all the rest of the many-celled animals.

In the evolution of the Metazoa one of the earliest 'inventions' must have been that of a mouth leading into a primitive digestive cavity.

This enabled Metazoa to tackle relatively large animals and plants as prey, whereas a sponge, by its whole construction, can never rise above microscopic particles—sifting the water for the sake of its debris as men sift rubbish-heaps for the few useful objects they may contain.

Metazoa, too, must have been fixed and sessile animals at the start, only later arriving at the emancipation of a freeswimming existence.

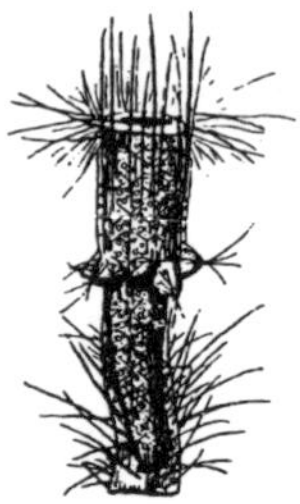

Fig. A Young Calcareous Sponge (*Sycon*)
Soon After Metamorphosis.

Their simplest representatives are all put in the phylum Coelenterata. A primitive coelenterate is essentially a small bag or tube, its walls made of two sheets of cells. It is fixed at one end, and has a mouth, surrounded by tentacles, at the other. The mouth leads into a cavity called the coelenteron, because it fulfils the functions both of the coelom and of the enteron or gut of higher forms.

Such a type is in the common freshwater polyp Hydra, which catches water-fleas and such-like (relatively) large prey as it droops from a water-plant with hanging net of extended tentacles. The prey is held and paralysed by a strange device. All over the body, and especially on the tentacles, are numbers of 'thread-cells', capable when stimulated of throwing out a hollow barbed thread containing poison. These thread-cells occur throughout the group; in some of the larger species, such as large jellyfish, they can inflict unpleasant damage on man, and even those of a sea-anemone can pierce our skin and make it tingle.

In yet another respect even the lowest coelenterates are more advanced than the sponges; they have muscles all along the body, so that the whole animal and not only isolated parts can be expanded or contracted. These muscles, however, are (at least in the lower coelenterates) of a very primitive nature, since they are only the inner ends of the epithelial cells forming the chief bulk of the two layers of the animal. Division of labour has not gone so far as in higher forms: the same cell contracts with one part of itself, protects or digests with the rest. So a village shop often performs post-office functions in one part, grocery functions in another, and stationery functions in a third, while in a town there will be separate shops for each function.

Most if not all of the group possess nerves, and at least scattered sense-organs for perceiving touch-stimuli. The most important contribution made by coelenterates to evolution was perhaps the first emancipation of Metazoa from a fixed existence on the bottom or attached to water-plants or floating objects to become free-swimming. Imagine a polyp like Hydra turned upside-down, the jelly between the two cell-layers much thickened, and the region between mouth and tentacles pulled out. This would give a fair idea of the way in which a typical free-swimming coelenterate is constructed. Such animals are called Medusae, or jelly-fish. While the smallest are microscopic, many are over a foot across, and a few are much larger still (up to nearly 8 ft. in Cyanea) and must weigh at least half a ton.

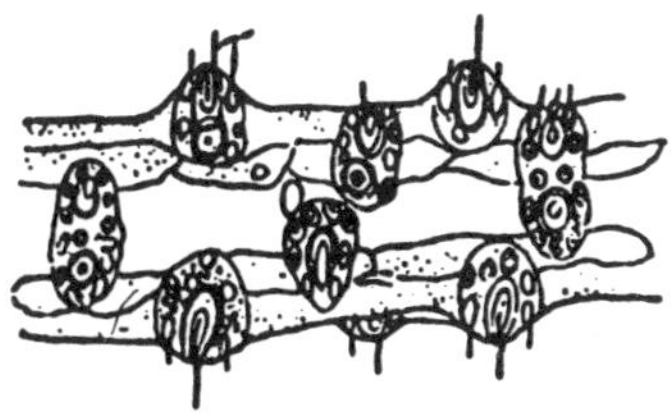

Fig. A portion of a Tentacle of Hydra

Fig. The aquatic larva of an insect after being captured by a Hydra

In spite of their size, however, their swimming is of a rudimentary kind. They are never swimming anywhere in particular, but drift near the surface of the sea; all their muscular energy is devoted to preventing themselves from sinking. Almost the only movement they can execute is the simple contraction of the bell, more strongly or less strongly according to circumstances.

Accordingly, they have no need of an elaborated nervous system. As a matter of fact, that which they and other coelenterates possess is of a primitive type known as a nerve-net, in which the sense-organs communicate with a network of nerve-cells branching all over the body, which in their turn communicate with the muscles.

The thick black lines represent the outer and inner margins of the body-wall. The nerve-net is figured in surface view over one tentacle (on the right), the base of the mouth, and the main part of the body and stalk.

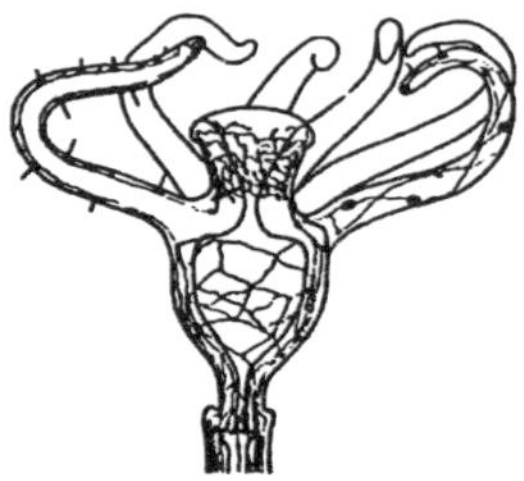

Fig. Diagram of the Nerve-net in a Hydroid Polyp.

The absence of any central nervous system, but the greater concentration of the net in the more sensitive mouth-region. The black dots on the nerve-net are bodies of nerve-cells. The hemispherical black bodies in the tentacles are nettle-cells.

Fig. A Hydroid Polyp (*Bougainvillea*).

The main nervous system of a vertebrate can be compared

with an elaborate telephone system. The nerve-net is somewhat like a telephone system in which every time any receiver was taken off its hook all the telephone bells of all the subscribers would start ringing.

This sounds inconvenient; but, as a matter of fact, as the 'subscribers' are the musclefibres, and as in jelly-fish they are all engaged in doing the same job in the same way, it is a good thing that all of them can be speeded-up or slowed-down by a single stimulus to one of the sense-organs.

In ourselves the intestine has to carry out movements of the same sort—the constant succession of peristaltic waves of contraction may be slowed, or accelerated, or in rare cases reversed in direction, but nothing more elaborate. And it, too, possesses a nerve-net not unlike that of the coelenterates. If one desires to visualize evolutionary progress, one may do worse than to remember that in its nervous and muscular organization a jelly-fish, for all its beauty, is pretty much on the level of the human gut.

Colony-formation is very frequent in coelenterates, largely as a consequence of the ease with which budding occurs. Corals are coelenterates and coral-reefs the accumulation of the skeletons of these little colonial polyps. Although there are very few coral-like animals round the English coasts to-day, in part of the secondary geological period they were abundant. Oxford was near the centre of a coral sea; and on the hills round the town one can dig out any quantity of fossilized corals.

Not only are some medusae very large, but the seaanemones, too, have found means to multiply very considerably the original insignificant size of the primitive polyp. This they have done by increasing the thickness of jelly between their two layers, and by dividing their enlarged central cavity by a series of strengthening partitions.

But both these and the jelly-fish seem to have been blind alleys for developing life. The main stream flowed elsewhere. All higher groups abandon the radial symmetry of the coelenterates and are (permanently or in early stages) bilaterally symmetrical, and therefore have a definite head-end and a back and belly.

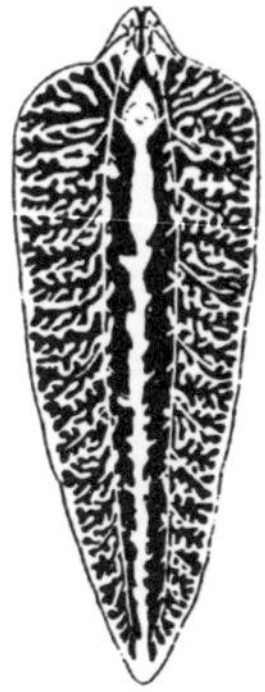

A. Digestive and Nervous Systems

B. Reproductive System

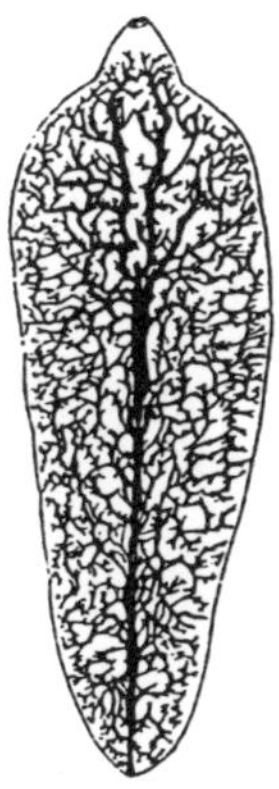

C. Much-branched Excretory System

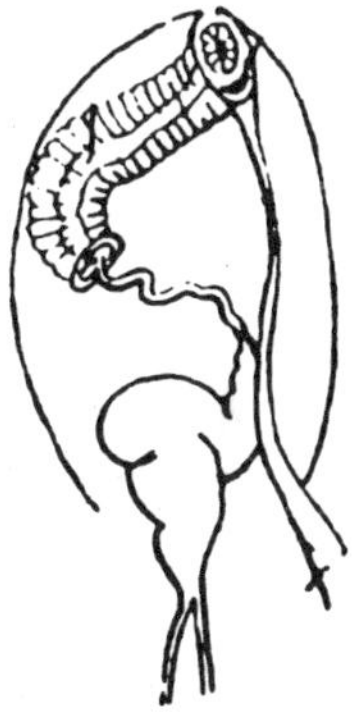

D. Enlarged view of region near genital aperture

Fig. The Anatomy of the Liver-fluke (*Distomum hepaticum*)

They also all early develop three main layers. The outer produces nervous system, sense-organs, and outer skin or epidermis; the inner produces the digestive tube and its appendages like digestive glands, &c.; the middle produces muscles, connective tissue, reproductive organs, blood-system, and, in vertebrates, kidneys and skeleton.

The lowest of these three-layered forms are the flatworms, with which we have already become familiar in the person of the Planarian. They possess no skeleton, no respiratory organs, no blood-system, no bodycavity, but only cellular 'packing' round the main organs, and only one aperture to the digestive cavity. Owing to the absence of blood-circulation, they must in the first place be flat and leaf-like, to put all tissues within range of oxygen diffusing from the surface.

In the second place, they must have their gut and all their organs finely branched. This is necessary so that the tissues of all organs may be able to acquire food by diffusion directly from the gut. Once a circulatory system was evolved, all necessity for this extraordinary branching of organs came to an end. The flatworms, however, show a great advance on the coelenterates in possessing a definite central nervous system with headganglia or primitive brain.

Space forbids more than a bare mention of two considerable other groups of worms, the round-worms or

Nematodes and the Nemertines. They are chiefly of interest to us in that they show stages in the development of new cavities in their bodies in addition to the digestive tube. The round-worms have a spacious cavity which probably corresponds to the blood-system of higher forms, although there is no true circulation in it, while the Nemertines show what is probably the beginning of the coelom.

In addition, their digestive tubes have acquired a second opening, the anus, at the opposite end to the mouth, so that the faeces can for the first time in Metazoa pass out at a different aperture from that at which the food is taken in.

The tiny rotifers or wheel-animalcules, familiar to all who use the microscope to investigate the population of freshwater ponds and ditches, are at about the same level of evolutionary advance as the round-worms.

They very much resemble the larvae of segmented worms and of molluscs, however, and are perhaps to be considered as animals which have become degenerate by never growing up but remaining permanently in an early stage of what was once a longer development. Save for a few exceptional cases, all the remaining members of the animal kingdom, which are sometimes grouped together as Coelomata, are characterized by the possession of three main layers—an anus, a true body-cavity or coelom, and a true blood-system.

The advantages accruing from these advances are clear. In the first place, the contraction of gut and muscular bodywall can now become more and more independent of each other, instead of the gut being squeezed or pulled out according to the movements of the whole animal. In the second place, a sort of trap is interposed between the gut and the rest of the body, in which poisons and actual bacteria passing in from the digestive tube can be dealt with.

In higher vertebrates patches of tissue which produce white bloodcorpuscles are found scattered over the inner wall of the coelom where it covers the intestine; while in many low forms such as worms, many white corpuscles laden with waste materials are found in the coelom, there to break down, and to be carried away by the excretory organs.

Thirdly, as the bulk of animals becomes greater, it becomes more and more necessary to provide greater absorptive surface in the intestine. We have already seen that in any structure which is enlarged without alteration of shape, bulk increases as r_3, but surface only as r_2. Thus, it will be of no avail to keep the same proportions of intestine as the animal grows larger, but new arrangements must be made for making the surface more or less proportional to r_3.

In some animals, like the earth-worm, this is accomplished by a simple infolding of one side of the straight gut; in others, like the sharks and dogfish, by a spiral valve in the intestine, down which the food must travel, as if down a shallow spiral staircase instead of dropping directly down a shaft. But in the majority of large animals the difficulty is surmounted by coiling the gut.

In a tadpole the intestine is packed like a watch-spring; even in man the gut is about four times as long as the whole body; while in herbivorous mammals it is often relatively much longer.

Only by the existence of a space such as the coelom would it be possible for the gut to become coiled in this way. The whole problem of size in animals is of great interest. The range of size in organisms is enormous, a big tree being as many times larger than a small bacterium as the sun is larger than the big tree.

It is startling to find that there exist adult insects, with wings and legs, compound eyes, and striated muscles, smaller than the human ovum; that jelly-fish may reach nearly a ton in weight; that the largest elephant has clearance top and bottom inside a whale; or that there are Protozoa larger than the smallest Vertebrates. Many problems as to the limitations of size both in an upward and a downward direction.

One of the most primitive groups of definitely coelomate animals is the Annelids, or segmented worms. These include the familiar earth-worms and their less familiar small freshwater relatives, the marine group called Polychaetes or 'many-bristled', on account of their numerous swimming or crawling bristles and spines on each segment (the lugworm is a familiar example); the Leeches, and some little-known aberrant types. One of the important characteristics which these possess is segmentation. Of the three highest phyla of animals, the Molluscs, Arthropods, and Vertebrates, it is no coincidence that two, and the two most successful, are segmented.

The meaning of segmentation is thus worth some study. All segmented forms are alike in certain respects. They all possess a small region in front of the mouth in which the brain, or its first evolutionary rudiment, is lodged. The trunk consists of a whole series, often a hundred or more, of segments, each one similar to all the rest in original plan.

These are formed during development in a growth-zone near the hind-end, and this growing-region often continues active throughout life. Thus, the trunk-region is, in a sense, repeated a number of times. Such segmentation is called *metameric*. What are the advantages of such a construction? They are several.

In the first place, the animal obtains any advantages of increased size that it would reap through colonyformation, but with the fundamental difference that the numerous identical parts, instead of being independent of and perhaps at cross-purposes with each other, are all under the control of the single anterior region, so that it is an organism of unified action from the first.

Then division of labour can step in, just as it can among the members of a colony, and modify different segments for different functions, so that high specialization is easily achieved. One of the earliest advances to be found in segmented animals is that of head-formation.

The primitive region in front of the mouth scarcely deserves the name of *head*. Gradually, however, several of the next following segments become firmly joined to it, their nerve-ganglia in particular all being fused to form a single brain of compound origin.

At the same time, more and more elaborate sense-organs, more and more elaborate jaws and mouth-parts, are evolved in connexion with what we can now call a *head*.

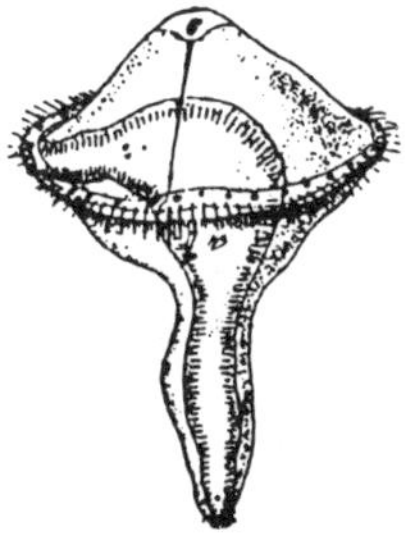

A. The Adult worm, dorsal view, tentacles, head, anus

B. Early larva (*trochophore*) eye-spot on sensory region

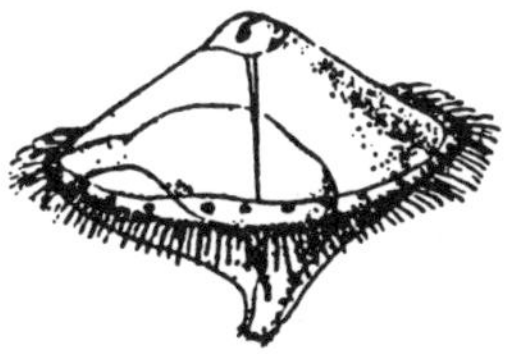

C. Late larva the trunk region has elongated, larval kidney

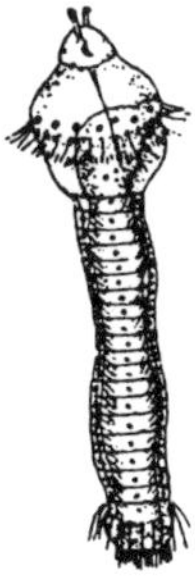

D. The Larva is Metamorphosing into the adult form

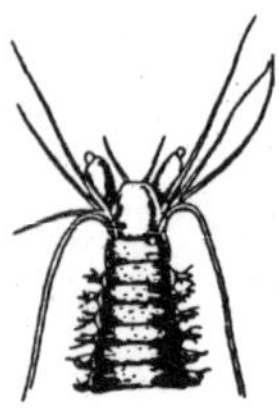

E. Anterior region of a Polychaete worm, the common sand-worm *Nereis*

Fig. Development, From the Early Larva onwards, of the Marine Annelid Worm Polygordius

The Annelid worms never get very far along this line. Their chief interest to the evolutionist lies perhaps in the presence in their most typical representatives of outgrowths all along the body, one to each segment, called *parapodia,* which one might translate as 'almost feet'. They consist of protuberances on either side of each segment, each furnished with a battery of bristles and hairs of various shapes. These sometimes serve for burrowing, sometimes for crawling, sometimes for swimming; there can be little doubt that from something of their type were evolved the limbs of crustaceans and insects.

Many worms have red blood, red with a haemoglobin similar to that found in our own veins—a good example of the unity underlying very diverse forms of life.

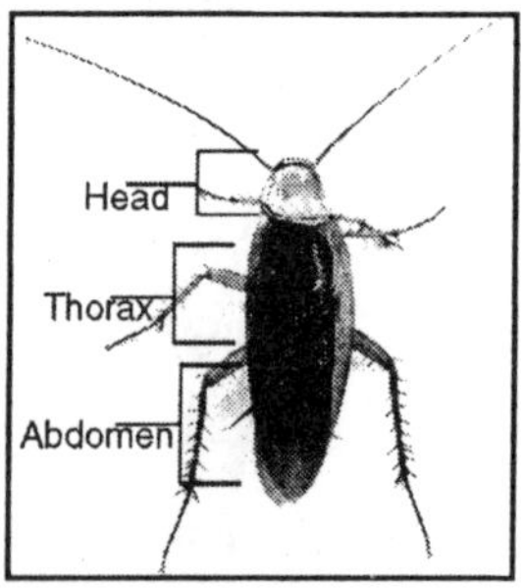

Fig. Side View of a Male Cockroach (*Periplaneta Orientalis*)

Others, however, have blood which is colourless, blue, or even green; in some cases the blood-pigment contains copper or zinc instead of iron. Worms play a great role in the soil; but that is a story every one should read for themselves in Darwin's book on earth-worms.

From some type like that seen in marine Annelids there probably sprang the great group of Arthropods—the largest group of the animal kingdom in respect of numbers, and in some ways the most specialized and even the most advanced.

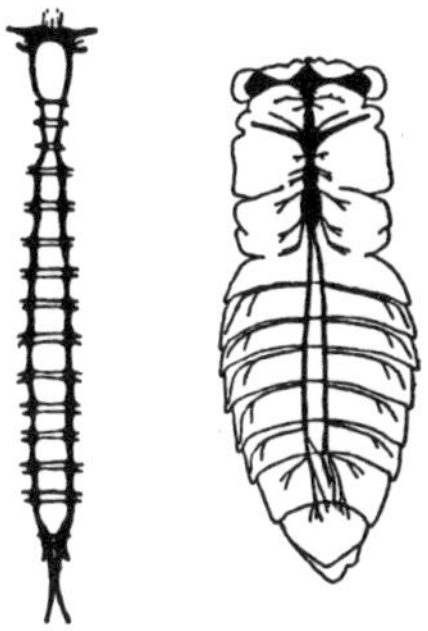

Fig. Concentration of the Nervous System in Arthropods

The decisive steps that they took in their evolution were these. In the first place they have all encased themselves in a hard covering made of a horny substance called chitin. This, by giving the possibility not only of unyielding attachment for muscles, but also of a fixed and definite shape, made rapid locomotion possible.

The primitive parapodia of the worms have been improved and converted into definite appendages, each consisting of a number of hinged joints. In all Arthropods, marked division of labour has set in among the appendages, so that between them they carry out at least three functions.

Some, like the antennae, have become senseorgans, the majority are used as limbs for walking or swimming; and the rest are modified into feeding-organs, taking the place of our lips, jaws, teeth, and tongue. In the higher members of the various Arthropod groups, the nerveganglia become concentrated near the anterior end, thus giving greater centralization of nervous control.

The number of segments may become fixed and definite,

while the head is always a sharply defined region. Curiously enough, neither cilia nor flagella, for which many uses are found elsewhere, from the lowest to the highest animals, are to be found in any Arthropod. The Nematodes are the only other group without cilia.

The most successful of aquatic Arthropods are the Crustacea, ranging from the little water-fleas and tiny forms like Calanus, which constitutes one of the main sources of food to many marine fish, up to lobsters and crabs. Shrimps, prawns, hermit-crabs, spider-crabs, wood-lice, sand-hoppers, crayfish, are all familiar members of this great group.

Although a few Protozoa and all earth-worms live in the soil and occasional flatworms and leeches are terrestrial, yet the Arthropods are the first phylum we have met, and the only one besides the molluscs and vertebrates, in which emancipation from a watery life has been achieved by the majority of whole classes or orders.

A few land-crabs are known; but insects, myriapods (centipedes), spiders, and scorpions are chiefly and typically land-dwellers, and it is especially among the insects that progress is most marked. Insects possess a remarkable method of breathing, wholly different from ours. Their whole body is penetrated by a network of air-tubes or tracheae. By this means oxygen and carbon dioxide are taken directly to and from the tissues, so that the blood has no concern with respiration, but only serves to transport food and waste-products. As a secondary consequence of this, the blood-circulation need not be, and is not, rapid; and the heart is of a comparatively low type.

Not content with conquering the land, the majority of insects are also at home in the air—a double achievement only found elsewhere among certain groups of vertebrates. Curiously enough, instead of using any of their existing limbs for flight, as have all flying vertebrates—whether flying-fish, flying-frogs, flying-lizards, pterodactyls, birds, or bats—they have employed as wings two pairs of quite new structures growing out from the upper part of the thorax, and probably developed from a kind of gill.

Every one has heard stories of the extraordinary capacities of various kinds of insects. One wasp has been seen to use a stone to pound down earth over its eggs—the only tooluser but man. The social life of bees and ants is more complicated than that of any other animal except ourselves. In a beehive, the newly emerged bees clean and prepare the cells to receive new eggs; after this, they help in keeping the temperature of the brood up when needful by clustering in dense masses over the nurseries.

A. The Main Trunks of the Tracheal System

B. Digestive and Reproductive Systems of a Female

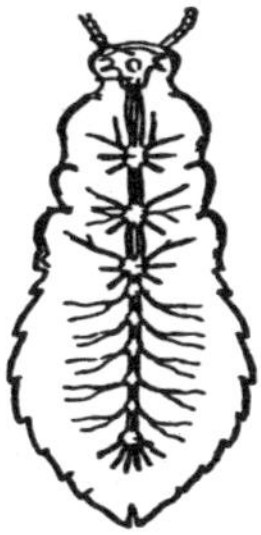

C. Central Nervous System

Fig. Internal Anatomy of the Cockroach.

They probably also ventilate the hive by fanning with their wings. When the workers are three days old they begin to act as nurses, feeding the grubs with honey and pollen; this nursing-duty is given up by two weeks old at latest. When they are between five and fifteen days old they take their first flights out of the hive, thus gradually gaining a knowledge of the surrounding country.

After this the young workers begin to collect food from the newly returned food-gatherers, and store it in the storage-cells of the comb. The growing workers soon add to these duties that of sanitary workers, keeping the hive clean and removing any corpses. The last task undertaken before going out food-gathering is that of sentry-duty; the sentries examine every bee that alights at the entrance, and attack all robbers or unwelcome strangers.

Fig. Interior of an Ant's Nest

Finally, when about three weeks old, the workers set out for their final task of gathering nectar and pollen. Thus, there is a wonderful division of labour or allotment of tasks within the bee-community, but the different jobs are not carried out by different worker-castes, as was at one time supposed, but are allotted to different periods in the life-history. There are ants that keep slaves some have gone so far as to lose the power of looking after themselves, and have to be kept clean and even fed by slaves of an alien species. Then there are ants which use their own babies, in the pupa stage, to build the nest; the pupae have an abundant and sticky saliva, a gang of ants squeeze threads of this from leaf to leaf held in place by another gang.

To Show The Way In Which the workers arrange the developing young according to their degree of development. In the top three chambers are very small larvae (grubs); in the fourth, full-grown larvae; and in the bottom chamber, pupae. The larvae and pupae are often erroneously called 'ants' eggs'.

Others have a caste of workers that gorge themselves with honey until their bodies are quite spherical, and then hang themselves up on the roofs of special 'cellars' against the winter. When food is short, these living store-casks are taken down and 'tapped' by the rest.

There are ants which keep domestic animals-the little aphids—which they tend and keep for the sake of their sweet secretion, and there are ants which practise agriculture. They make subterranean hotbeds with pieces of leaves which they cut, and in these plant the spores of special fungi, sometimes only known in the ants' nests. When a queen goes out to found a new colony, she takes a pellet of the precious fungus with her in a special pocket below the mouth.

Fig. Two Gangs of Workers of the Ant Oecophylla Smaragdina repairing a rent in the nest

Insects appear to have been in existence for a longer time than vertebrates; certainly the highest insects such as ants have been in existence longer than the highest mammals. Why is it, we may well ask, that the vertebrates ever got the chance of rising? Why did not the insects come to occupy a predominant position among animals, and keep out intruders from their preserves?

Fig. A Store Chamber of the Honey-pot Ant (*Myrmecocystus*)

Certain workers gorge themselves with honey until their abdomens are quite spherical. They then hang themselves up in special underground chambers until there is a shortage of food, when they regurgitate their honey to the other workers. The answer appears to be twofold. In the first place, their cleverness and efficiency is far more a matter of inborn instincts than of learning, clockwork smoothness rather than discovery and choice.

They can learn, but the power of learning is small, the elaboration and fixity of their instincts great. It is precisely the reversal of this relation between instinct and learning capacity which finally enabled the vertebrates, in the person of man, to begin to rise above and to control the forces which up till then had controlled and moulded life.

Secondly, as it turned out, the typical structure adopted by the group carried irrevocably, within itself, a limit to any great advance in size. The skeleton of an Arthropod is not only on the outside; once it is laid down it is dead. If growth is to take place, the animal must moult—the old skeleton be split and thrown off and a new one formed underneath. While the new one is being formed the animal is naturally soft and defenceless. As an Arthropod increases in size, for various mechanical reasons the bulk of the skeleton must grow not only absolutely but relatively bigger. The difficulty of emerging from the armour-plating becomes greater, and the time necessary for building up a new skeleton and hardening it with lime would be more and more prolonged.

If there could exist a crab as big as a cow, it would have to spend more than half of its existence in hiding, waiting for its skeleton to grow after moulting. But such an animal could not exist. Even in water and without a skeleton a crab's body has some weight, and a crab of this size would at moulting flatten out like a gigantic bun.

This would happen of course at a much lower size in land Arthropods, whose weight is supported by no circumambient water. And as a matter of fact, although we find moderately large marine Arthropods, such as the giant spider-crab, with a body a little bigger than a man's head and gigantic but thin legs, yet the largest purely land forms (excluding land-crabs) are the big tarantula spiders, the bigger among the scorpions, the goliath beetles, and the Giant Swift-moth, none of which have a body (excluding limbs, wings, and the tail in scorpions) of over six inches long.

Fig. Below, an Adult Dragon-fly (*Aeschna cyanea*) soon after Emergence

The two very similar pairs of wings are drying in a vertical position. When dry, they are held horizontally. Note the huge eyes, the long segmented abdomen, and the three very similar pairs of legs arising from the 3-segmented thorax. To the left, part of the larval skin from which it has emerged. Above, the same specimen emerging from the nymph (*larva*). The wings are still very small; they are later dilated by the fluid pressure of the blood. In Dragon-flies the larva is aquatic, but not otherwise markedly different from the adult except in being incapable of flight. The wings are formed as external buds, whose skin is seen on the skin of the larval thorax. There is no resting-stage (*pupa*) between larval and adult stages. At this, and every moult of an insect, a complete new chitinous external skeleton is produced.

In order to obviate some of the difficulties which at each moult beset a complicated animal with external skeleton, in all the more specialized insects is found the plan of dividing life into two wholly distinct phases, with a metamorphosis between. In the first or larval phase the necessary growth is achieved, and the animal is very little else but a feeding machine—a butterfly's caterpillar, beetle's grub, or fly's maggot. Then comes transformation to a resting-stage or pupa, during which the metamorphic transformation is accomplished—the white blood-cells break down the larval organs, and little reserve packets of cells grow up into the organs of the adult. Then from the pupa there emerges the reproductive adult or imago, capable generally of flight, active, endowed with efficient sense-organs and wonderful instincts. In other groups there is no pupa stage, and the metamorphosis is less radical.

Thus, in insects the larval stage is a new development, forced upon the higher members of the group by their very complexity, not as in Amphibia a primitive condition, the larval amphibian to-day living in the same way and in the same element as did the ancestors of the group in the remote past. Two matters must detain us for a moment. The first is the social life of many insects. The bees, the wasps, the ants, and the termites all display a wonderful perfection of community life.

Well-developed community life with organized societies only exists in the Arthropods and Vertebrates. These are two groups in which the formation of colonies with physically connected members, so frequent in lower forms, is absent. The meaning of this is simple. The animal community, like the colony, increases the size of the effective unit; it takes the place of the colony in groups above a certain level of complexity; and this for two chief reasons. First, it would be of no advantage, but of definite disadvantage, to have, in any animals so highly organized as insects or vertebrates, a colony composed of individuals joined to each other by physical connexions.

A high type of animal is a high type largely by virtue of its elaborate organization for moving from place to place and for perceiving distant objects, and all this would be nullified by such an arrangement. On the other hand, the perfection of its sense-organs and its brain enables it to communicate with its fellows and to possess instincts making for concerted action. In a colony of polyps, the only way in which one polyp can help another is digestively; ants, however, will guide others to food they have found, and the adults will tend the larvae.

A community, in fact, is a colony held together by psychical instead of by physical bonds. The largest communities in the world are those of ants and of men. Some ant communities contain over half a million individuals. The largest human communities at present are the British Empire with about 460 million individuals, and China with about 420 millions. In all social insects the great majority of individuals are 'workers', incapable of reproduction. In ants, bees, and wasps the workers are modified females, while in termites both males and females have become unsexed. In many ants and termites the workers have become differentiated into several sub-castes, of which the largest act as soldiers.

The other point concerns the senses of Arthropods. Arthropods have progressed far in evolution, but along wholly different lines from those along which the vertebrates have travelled. As one would expect, they possess the same general kind of sense-organs as vertebrates—for touch, smell, taste, sight, and sometimes hearing—but these are often constructed on a plan wholly different from that of ours.

The sense of smell in insects, for instance, seems to depend upon the sensitive hairs in the feelers or antennae. These are very highly developed in the males of some moths, such as the Oak-eggar, and enable them to find a female of the species, even if shut up in a box, at a distance of a mile or more. The Arthropod compound eye, however, is perhaps the most remarkable of their sense-organs. It consists essentially of a number of little elongated eyes placed side by side, and separated optically from each other by a black backing of pigment to each.

The retina of each eye can only receive an impression of the tiny area of outside world directly in line with it beyond the window of its cornea. All the retinae transmit their impressions to the brain, which must then receive a mosaic of separate images, and must then combine them to a single coherent picture. In some dragon-flies the huge compound eyes are composed of over 25,000 separate eyelets, and are curved so that some of these point in every direction. A dragon-fly can look in front, and backwards, and sideways, and up, and down, all at once and without a movement of its head. Such eyes are probably more efficient than ours in some ways, such as in this matter of looking many ways at once, and in the detection of small movements of objects; they are, however, certainly less efficient in giving an accurate picture of the details and fine texture of objects, since our sensory units are smaller, and they have no focusing mechanism.

Insects, as we have seen, are usually small. Every one will have noticed flies and small beetles and moths struggling on the surface of water; they are so small that they are a prey to the surface-tension of the surface-film, and cannot get free. It is thus difficult for most insects to go to water to drink, like a land vertebrate (collectors of butterflies will have seen 'Blues' and other butterflies drinking—but always from moist soil, never from pools of water).

It is possible that an unusual anatomical feature of theirs is to be associated with this fact. Their excretory organs do not open directly to the outside, but take the form of a number of thin tubes (Malpighian tubules) opening into the intestine. This may possibly be a water-saving device. The hinder part of the intestine in man absorbs water; if it did the same in insects, all the water contained in the excretory fluid would be reabsorbed, and the animals would have to drink much less.

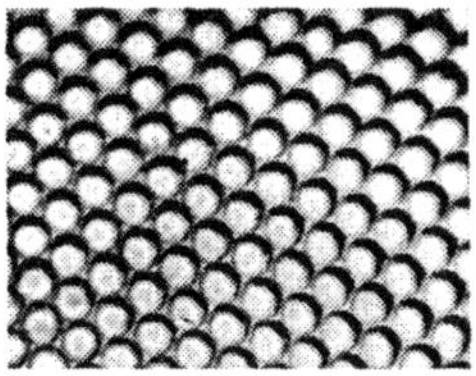

Fig. Micro-photograph of Part of the Surface of An Insect's Eye. The Large Number Of Separate Facets, Each with its own cornea, is clearly shown.

Of other Arthropods we can speak but little. Mention should be made of the wonderful crab Birgus, one of the few land-living Crustacea, whose claws are so powerful that they can open coco-nuts and even cut barbed wire; and then there are the spiders, which have developed a tracheal breathing system quite independently of the insects, and have the best-developed courtship displays of any invertebrates, as well as rivalling all but the social insects in complexity of instincts.

There remain of the invertebrates the Molluscs, the Brachiopods, and the Echinoderms. The molluscs comprise the true 'shell-fish', and are biologically an extremely heterogeneous group, ranging from tiny primitive worm-like creatures to the largest and most highly organized of all invertebrates—the squids and octopuses.

They all possess (besides other technical characters) a large fleshy organ of locomotion, the foot; and almost all have a shell; they differ from most other successful invertebrates in not being segmented. The most important are the 'Bivalves' (Lamellibranchs), and 'Univalves' (Gastropods), and the octopus, the nautilus, and their relatives (Cephalopods).

The bivalve molluscs are a degenerate group; their degeneration has come about through their adoption of sessile or semi-sessile habits. In practically all of them, such as the oyster, the mussel, the freshwater mussel, &c., the so-called 'gills', though they still assist in respiration, are in reality mainly food-catching organs. They are very large and covered with cilia; the cilia produce a current, and the gills are so arranged as to strain off all small debris from this current.

Their digestive system is so constructed that it cannot avail itself of large particles of food, and all particles above a certain size are side-tracked and ejected again if they manage to get in. The acquisition of food by these means does not require great intelligence or elaborate senses; and we find brain and distance-receptors very poorly developed in almost all members of the group.

The Gastropods are so called because 'They travel about on their tummy'. The most remarkable fact about them is that they are not symmetrical; usually, as in the common snail, a large part of the body is twisted up into a spiral, the spiral being covered with a shell.

Limpets, cowries, whelks, slugs, snails, periwinkles, sea-slugs, the pteropods that form the staple diet of whalebone whales—these are typical examples. They represent somewhat of a compromise in evolution between security and progress.

Fig. A Common Octopus (Octopus Vulgaris)

The foot of these animals is prolonged into eight 'arms' provided with suckers. The mouth, with horny jaws like a parrot's beak reversed, is between the arms. The right eye is seen, also the funnel from which water is ejected from the mantle-cavity.

Water is drawn in round the mantle-edge to aerate the blood in the gills. The animal can swim rapidly backwards by ejecting water violently through this funnel. It also crawls about by means of its arms. The body, containing the viscera, is on the left.

The shell is a fine protection—but it is a very heavy piece of armour-plate. As a result, we usually find movement effected by simple crawling on the slimy under-foot; the snail is not only proverbial but typical in respect of its slowness. Since they are free-moving, distancereceptors are wanted; but since they move slowly, these sense-organs are only moderately well-developed. Most of the species are neither very small nor very big—from about a centimetre to a decimetre long; a few, however, reach greater sizes; one fossil tower-shell stands nearly five feet high.

The highest molluscs, however, and in some ways the highest invertebrates, are the Cephalopods. The more primitive among them, such as the Nautilus and the Ammonites (now extinct, but once the dominant group), still retain a large protective shell. But in all the highest forms the shell is quite internal and much reduced transformed from external armour to internal support or even absent: the 'foot' has in these creatures become divided into eight or ten 'arms' beset with suckers.

In some of these higher forms an unusual method of rapidly moving from place to place has been evolved; they squirt out water forwards through a narrow jet, and so move quickly backwards.

In others, muscular fins are developed along the sides, by means of which the animal can swim either forwards or backwards. In addition some, such as the octopus, can clamber about on the bottom by the aid of their arms.

All are freely and often rapidly mobile, and not stuck to the bottom like most Gastropods. This has resulted in a great advance in eyes and other sense-organs, and in brain. Further, the division of the foot into lobes ('arms') has provided them with a different type of organ from any other seen in other molluscs—a real set of limbs. Finally, most of them are fairly large; some indeed are gigantic, thirty feet or more across the outspread arms.

These giants are for the most part inhabitants of deep water, whither they are pursued and attacked by the sperm whale. For all their complexity of organization, however, they are not a large or dominant group like the fish; something of the combined simplicity and efficiency of the fish is lacking in them, and they remain, like monuments of another type of civilization, to show the utmost that life was capable of producing along the Molluscan line of evolution.

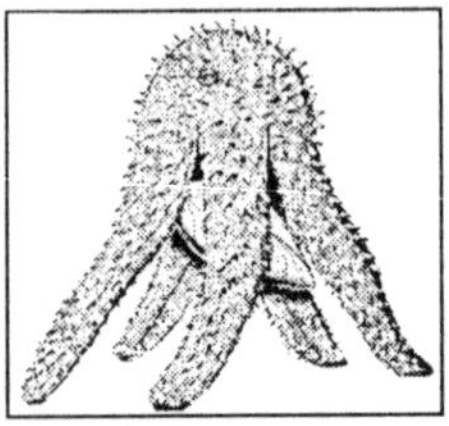

Fig. A Starfish (*Echinaster*) Devouring A Mussel (*Mytilus*)

Madreporic plate, through which water enters into the system of water-vessels, which help actuate the tube-feet by keeping them distended. By means of its tubefeet the starfish exerts a constant pull which eventually wears out the muscular resistance of the mussel. The starfish then protrudes its stomach out of its mouth and inside the shell of the mussel. Here it apparently digests the body of the mussel, the resultant fluid being sucked up into the rest of the digestive tube of the starfish.

The Brachiopods or lamp-shells are often mistaken for bivalve molluscs. In reality they have quite another type of anatomy, with coiled tentacles instead of gills for the production of their food-current. They are interesting because some of their species have persisted without the least visible change from the earliest fossil-bearing rocks till to-day. Evolutionary change is always occurring, especially among the latest products of evolution; but it is not a necessity. An animal may perform the same job in the world's economy for as long as the world is habitable.

The Echinoderms are a curious phylum. They are all marine and essentially bottom-livers, sometimes fixed by a stalk like the sea-lilies, more often capable of slow movement like the starfish, brittle-stars, sea-urchins, and sea-cucumbers. They are remarkable in being the only whole group of coelomates which have reverted from bilateral to radial symmetry. They are all five-rayed; but bear, in the form of a few small structures, unmistakable signs of an original bilaterality. As would be expected, their sense-organs are very poorly developed. Two very curious features may be noted. They move by means of a system of tube-feet—a great number of little protrusible suckers filled with water which can be made to adhere like a boy's 'sucker' by the creation of a partial vacuum. And some of them, such as starfishes, sea-urchins, possess the oddest organs, called pedicellariae, which every one who has a lens or microscope should examine when they are at the seaside.

They are little stalked structures, each with three 'jaws', and are continually moving from side to side making snapping movements, each one apparently quite on its own. It is probable that they help to defend the animal, and also to keep it from being overgrown with plants or sessile animals. Most Echinoderms have a larval stage, during which they are tiny transparent creatures, bilaterally symmetrical, swimming near the surface of the sea. It is clear that the Echinoderms represent the end of an evolutionary blind alley.

Chapter 12

The Vertebrates

These claim our interest not only as the group of animals with the highest average attainment, not only as the group which contains our own evolutionary pedigree, but because a great deal of the detailed course of their evolution can be traced in the fossil-bearing rocks. All the invertebrate phyla and most of their classes had their origin so far back as to antedate the first fossil-bearing rocks we know.

The earlier stratified rocks, that were laid down when they were first evolving, have been denuded away or so squeezed and baked through heat and pressure that their whole character has been altered, and the fossils they must have contained have been destroyed or rendered unrecognizable. Worms, echinoderms, arthropods, molluscs, corals, lamp-shells, many highly specialized and none essentially unlike those of today, are to be met with in the earliest well-preserved series of rocks, the Cambrian.

But with the vertebrates it is different. Probably their evolution took longer on account of their very complexity; in any case, the first vertebrates so far found belong to a primitive type of fish, and occur in the Ordovician, the next division above (more recent than) the Cambrian. The main features of vertebrate evolution could be deduced equally well from comparative study of present-day forms, or from the anatomy and history of fossils. The two methods confirm each other in all essentials. Starting with fish, the salient steps in vertebrate evolution are as follows:

- The partial conquest of the land by Amphibians, involving the transformation of swim-bladder to lungs, and of paired fins to true limbs with fingers and toes.

- The full conquest of the land by reptiles, no longer restricted to dampness when adult or to water for their early development. This implied the evolution of a large-yolked egg, and the development of a protective water-cushion or amnion over the embryo within the egg. Instead of having to develop in water, each embryo is supplied with what an American writer calls 'its own private pond' in the shape of the fluid within its amnion.

In higher reptiles (mostly now extinct and supplanted by mammals) the body was for the first time raised off the ground and supported entirely by the limbs (or with aid from the tail). Meanwhile, the heart became more or less completely divided into two separate parts, as in man, one for pumping venous and the other for pumping arterial blood only.

- Two separate lines spring from the reptilian stock—the mammals and the birds. Both agree in having developed a mechanism for ensuring a constant temperature-environment for the tissues of the body, and in having the heart completely divided into two. The first to be considered here (though the later to develop in evolutionary time) is the bird line. The evolution of birds was made possible by a series of acquisitions. First, that of constant high temperature, next those of feathers, wings, and air-sacs. In addition, existing birds have lost their teeth, and birdparents show a remarkable degree of care for their young. These steps have led to the chief conquest of the air which has been made by vertebrates.
- The other line led to the mammals. Apart from the constant high temperature and the divided heart, the two universal characters of mammals are the possession of hair, and the secretion of milk by the mother. Further, although a few mammals lay eggs, and a moderate number (most pouched mammals or

Marsupials) bring their young up from an extremely early stage in their pouch, stuck firmly on to the nipples, yet the largest and the dominant sub-class of mammals are all characterized by a placenta or organ for ensuring interchange of food, respiratory gases, &c., between the mother and the embryo in the uterus, thus making possible not only a speedier development but also the protection of the embryo within the mother until a later stage than occurs anywhere else in the animal kingdom. All typical mammals are also characterized by having a division of labour among their teeth (incisors, canines, and grinders), which only occurs elsewhere in one extinct group of reptiles.

One or two interesting points emerge. The fishes have continued their unabated success, as the most generally successful group of water-living animals, from very early times up till the present: they hardly compete with mammals or birds. But the Amphibia had their hey-day, and sank with the rise of Reptiles; the Reptiles had a still more marked and more remarkable period of dominance, ending with reptilian collapse and avian and mammalian advance, and the non-human mammals are now showing the same kind of decrease coincident with the rise of man. Thus in the fossil record a succession of types is really visible, and the succession is definitely one of lower by higher types.

There are, however, other Chordates beside these five classes. Space forbids mention of some of the doubtful 'poor relations' of the stock, and only allows the briefest reference to the Tunicates. These latter are a degenerate group which have lost many of their distinctively Chordate characters. Their degeneration, as in the bivalve molluscs, is due to their having adopted the method of feeding by producing currents and straining off the food-particles; this has led to a sessile mode of life and to loss of sense-organs and diminution of brain.

Many of them form colonies by budding, and have surprising powers of regeneration and dedifferentiation; they are hermaphrodite. They have also lost their skeleton; indeed, each one of them possesses it as a free-swimming larva, and loses it and all the main senseorgans when it metamorphoses and settles down. They must have branched off, however, at a very early period from the main Chordate stem; and it is perhaps comforting to reflect that if they have lost their skeleton, it was by then only a notochord and not a real backbone. A related form, but one much closer to the original primitive Chordate type, is seen in Amphioxus.

A. The free-swimming 'tadpole' larva

B. The larva has fixed itself

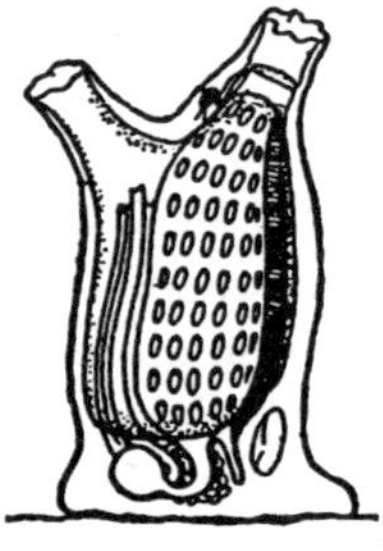

C. Metamorphosis is complete

Fig. The Metamorphosis of A Sea-squirt (*Ascidian*).

The most important things about Amphioxus are those which it does not possess. It has nothing that could be called a head; the merest apology for brain and distance-receptors; no skeleton except a notochord; no heart, but only an ordinary bloodvessel which contracts rhythmically, a liver which is a mere unbranched pocket of the gut, no limbs, no reproductive ducts, the eggs and sperm simply bursting out through the body-wall. It is, however, without a question a Chordate, as shown by its notochord, its pharynx pierced by gill-slits, and its hollow nerve-cord running along the back instead of the belly. It too depends on artificial currents for its food.

It serves to remind us of a time long before that of the earliest fossils preserved to us now, when none of the Chordate stock had reached a higher organization than this; and all the highest types of life—bird, horse, lion, dog, and man himself —were no more than a potentiality slumbering in the germcells of little Amphioxus-like creatures in the sea.

The next stage in vertebrate evolution of which we have any record is represented by the Lampreys; but there is an enormous gap between them and Amphioxus. The Lamprey has already a well-developed brain, a rudimentary skull and backbone of cartilage as well as a notochord, 'nose', eyes, and ears, and a proper vertebrate heart and liver. But they are still far behind any true fish. They have no limbs, no true jaws, no true teeth, and none of them have bone. They, too, like Amphioxus and its relatives, make but a very small group, a relic from the past.

In their life-history, they shed a most interesting light on the evolution of the thyroid gland. The lampern, as the Lamprey's larva is called, still obtains its food from a foodcurrent, in the same way as Amphioxus and the Tunicates. One of the special features of the straining mechanism of all these Chordates is a groove called the endostyle running along the floor of the pharynx, which secretes a sticky mucus.

This is forced forward by cilia, round the mouth in two grooves, and along the dorsal groove back into the intestine. The gill cilia are so arranged that all food-particles strained off by the gills are driven up to the dorsal groove. Here they become entangled and stuck in the slime, and are passed on by this sort of moving stair-case to be digested in the intestine.

The lampern has an endostyle just like that of Amphioxus, except that it is rolled up in a sort of pocket under the pharynx, and sends out its slime-cord ready made.

A. A Piece of skin, slightly magnified, in surface view. It is covered with denticles, their points projecting obliquely upwards.

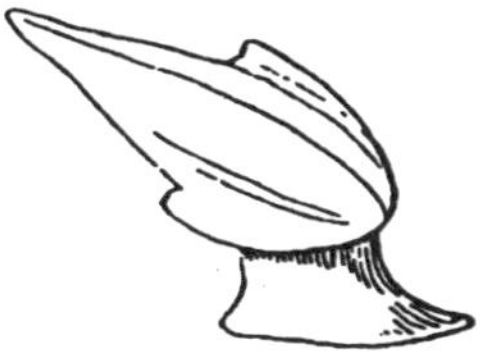

B. A Single denticle. *b*, the portion embedded in the skin.

C. A Denticle in Section.

Fig. Denticles of the Dogfish

When the lampern changes into the Lamprey, most of the endostyle degenerates altogether. But some of the cells of its duct remain alive, multiply, and become converted into a typical thyroid. This is a transformation that nobody would have been rash enough to guess at, if they had not been able to see it actually happen, and it is of interest as showing the way in which one organ, no longer required by the animal, may become converted into another organ instead of disappearing altogether.

The balancing planes—the paired fins—of fishes become supporting limbs in land forms; the swim-bladder which regulates a fish's density and consequent distance from the surface becomes a breathing organ—the lung; hair in man has lost its primitive warmth-retaining function, and under the influences of sexual selection has become converted to an adornment. A gap again yawns between the Lamprey and the Fish, although not such a wide one as that between Amphioxus and Lamprey. No fish has a notochord persisting at full size throughout life; all have well-developed vertebrae and skull over-arching the brain, paired limbs, scales, and teeth.

The true jaws have appeared, and can be shown to have arisen by another strange change of function; they are derived from the first pair of the bars of cartilage which support the gills and hold the pharynx-cavity stretched as an open umbrella is held out by its ribs. In almost all fish, however, the upper jaws are not yet firmly united to the skull as in all land forms, but only jointed on. Teeth, too, have an odd history. The skin of dogfishes and sharks can be used for polishing, and when prepared is known as shagreen. Its qualities are due to thousands of little pointed scales sticking out from its surface. When these are examined, each is seen to be nothing else but a miniature tooth fixed by an enlarged base.

Before teeth served as teeth they were scales, and covered the whole surface of the body. It was those in the skin covering the jaws which were able to take on new functions, became true teeth, and eventually alone remained when all traces of the skin-denticles had disappeared.

There are two main groups of fish—the Elasmobranchs (dogfish, sharks, skates, and rays) which have never developed bone in their skeleton, and the Teleosts or higher bony fish, comprising all the familiar species like herring, sole, trout, cod, sea-horse, and flying-fish. The former are primitive in a great many ways. In one respect, however, they are better equipped than the Teleosts.

They lay a few largeyolked eggs, well protected in horny capsules (the 'Mermaids' Purses' one picks up on the seashore) or even in the oviduct of the mother; while the latter lay their eggs before fertilization, and have to produce vast quantities of them in order to compensate for the inevitable wastage during their tiny and unprotected early lives. The cranial nerves are marked with roman numerals. The parts of the brain concerned with various senses are marked as follows: smellcoarse dots; sight crosses; hearing, balance, and lateral line organs-broken oblique lines; touch and other skin senses—vertical lines; taste and other stimuli from viscera (gills, stomach, etc.) horizontal lines.

The motor nerves to gills, stomach, and other viscera are marked in black and white rectangles. It is probable that this one specialization enables the Elasmobranchs to compete not too unsuccessfully with the Teleosts. Their brain is very primitive, even when compared with that of the frog. The smallest fish are under an inch in length, the largest is a form of shark which may reach forty feet.

Fish have evolved into the most extraordinary forms. 'Funny Fish'—pipe-fish, sea-horses, ribbon-fish, sun-fish without a proper tail, cowfish, flying-fish, parrot-fish, porcupine-fish, electric-fish, the flat-fish which fall over on one side as they grow and twist both eyes over on to one side of the head, the remora with a sucker on its head to attach itself to its living locomotive the shark, angler-fish, deep-sea fish with eyes on long movable stalks, or with mouth as big as the whole body, or with rows of red and white phosphorescent lights like a liner at sea.

Fish are the dominant group in the waters, and have become specialized to fill every available niche. One group of freshwater fish, the lung-fishes, are able to survive a long sojourn in the mud when the ponds and streams dry up. This they do by extracting oxygen from the air taken into their swim-bladder, which thus acts as a lung when needed. Thus, these creatures partly bridge the gap to terrestrial life.

In the structure of the limbs, however, a great gap exists between fish and amphibians. No fish has anything but fins, no amphibian anything but legs equipped with fingers and toes. Again, just as the thyroid gland was evolved from the remains of the endostyle when this ceased to be of use, so another ductless gland, the parathyroid, is not found in fish, but only arises out of the debris of the gill-slits when the vertebrates took to land. The Amphibia need not detain us long. They are in a certain sense a compromise between life in water and life on land. The biggest existing amphibian is the giant salamander, which may reach four feet, though the Amphibia as a group average six inches or less. However, in the Carboniferous period, when they were the only land vertebrates, the average was much higher, and forms over six feet long existed; but all these disappeared as soon as the reptiles entered the field. One other point deserves mention. The Amphibia are the first vertebrates capable of producing vocal sounds deliberately. The reptiles were the true conquerors of the land.

Ceratodus, with well-developed fins, from Queensland.

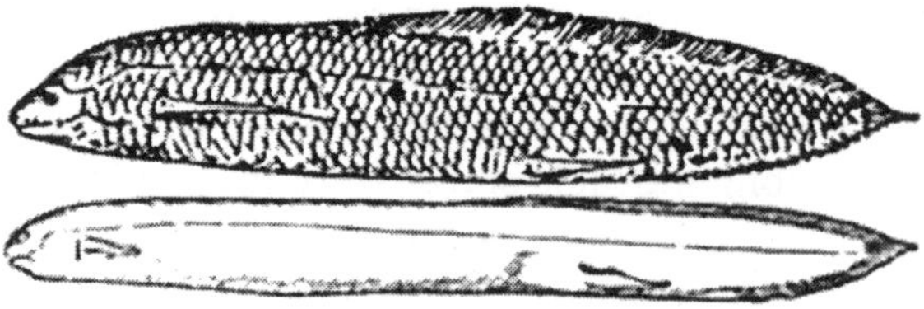

Protopterus and Lepidosiren, from tropical Africa and South America Respectively, with reduced fins.

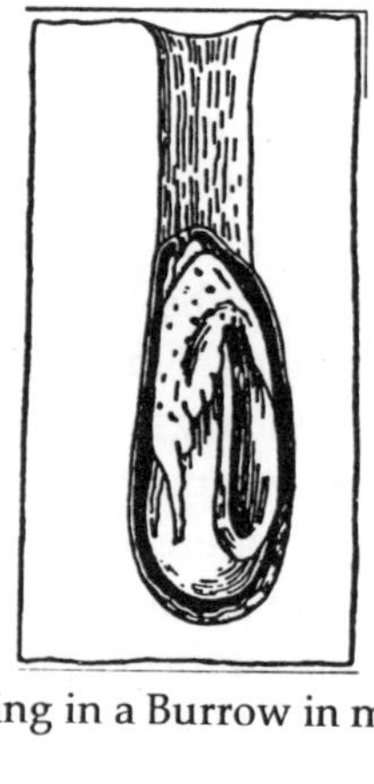

Protopterus Aestivating in a Burrow in mud in the dry Season

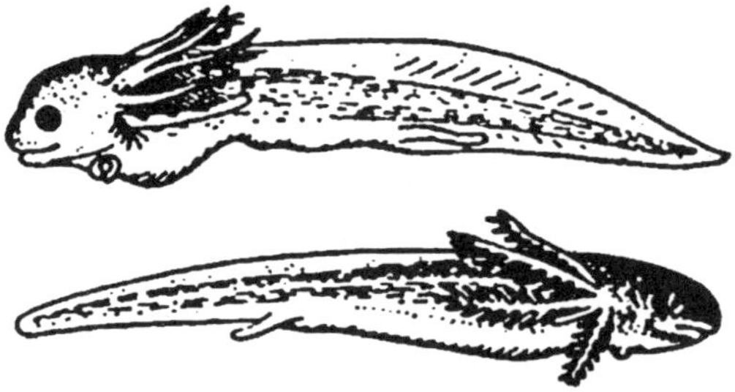

Larvae of Protopterus and of Lepidosiren

Fig. Various Lung-fishes (*Dipnoi*)

This conquest they owe chiefly to their dry, strong skin, and the evolution of special membranes helping the embryo to live away from water in a large-yolked egg. The allantois, from which the placenta afterwards developed, is the embryo's breathing organ: the amnion is a protective water-cushion, enabling the soft embryo to develop in fluid, protected from pressure and contact.

In their hey-day the reptiles rivalled the present mammals in size and variety of specialization. Besides the existing lizards, snakes, crocodiles, and tortoises, there lived in the middle and late secondary period a whole series of remarkable types. There were mammal-like reptiles, equipped with several different kinds of teeth, and able to run like a typical quadruped; flying pterodactyls; at least two types which had gone back to the sea, the icthyosaurs, which more or less resembled whales (and produced their young alive, as a specimen in the South Kensington Museum testifies, with a brood of embryo skeletons between its ribs), and plesiosaurs, with great flexible necks, creatures which must have looked very much like the average man's idea of the sea-serpent; and C finally, the most successful group of all, the dinosaurs, including rapid runners on two legs, living 'tanks' covered with armour-plate, great semi-aquatic herbivores like Diplodocus , some of which grew to a hundred feet long, and the biggest carnivorous creatures ever known, such as the Tyrannosaurus, which stood over twenty feet high, and no doubt lived upon the gigantic vegetarians.

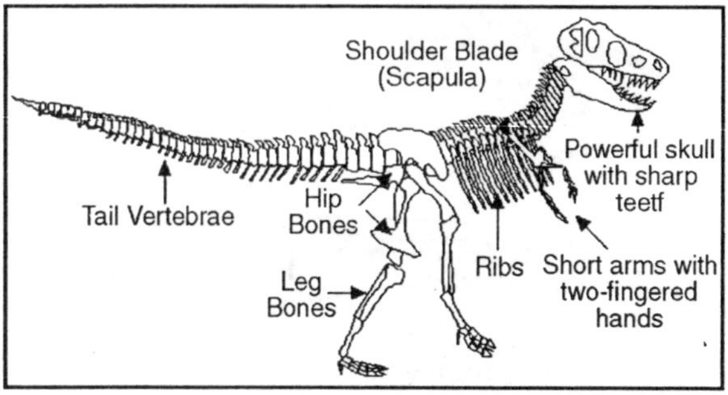

Fig. Skeletons of the Extinct Dinosaur

The end of the Secondary period comes, and with the beginning of the Tertiary the pride of the reptiles is humbled. More than half the groups, and those the most advanced, no longer exist; those that are left are already playing second fiddle to the early mammals and birds.

What brought about this revolution is not certain. Possibly an alteration of climate cut down the available food-supply and gave advantages to smaller creatures capable of temperature-regulation. There can at least be no doubt that temperature-regulation and better provision for the young, both before and after birth or hatching, are the two progressive features in which birds and mammals chiefly outdistance their ancestors, the reptiles.

In the case of birds we luckily have come to possess true 'missing links' between reptiles and modern birds. In the mid-secondary, there lived a creature called Archaeopteryx-'Earliest winged creature'. It was an undoubted bird, for it possessed feathers and obvious wings. But its jaws possess a good complement of teeth, the wing is still extremely primitive in possessing claws on three of its fingers (by means of which it no doubt crawled like a bat among the branches), and its tail is not a fan as in all living birds, but is more like that of a kite, with a long jointed skeleton, and feathers coming off on either side all the way down.

Fossil birds found in strata from the close of the Secondary already possessed the modern fan-like tail, and had lost the claws on the wing; but they all still possessed teeth. Toothless birds only appear with the Tertiary period. Flight, high temperature (over 100°, sometimes as much as 105°F), air-sacs and hollow bones, nest-building, bright colours and elaborate courtship, song, and the care of the young—these are the chief characters of modern birds. They owe their success chiefly to one single character—the evolution of feathers. These in the first place keep down radiation and so allow of a high body-temperature.

They also permit of the fore-limbs alone being used in flight. Thus the hindlimbs are left free to develop along their own lines, instead of being used up, so to speak, as one of the supports for a wing-membrane, as occurred in the extinct flying lizards, and occurs to-day in the bats. The air-sacs not only lighten the body and help in breathing, but are used to stream-line the body so that it offers least possible resistance to rapid passage through the air. In the same way, most of the fuselage of an aeroplane only serves for stream-lining, not for carrying passengers or goods.

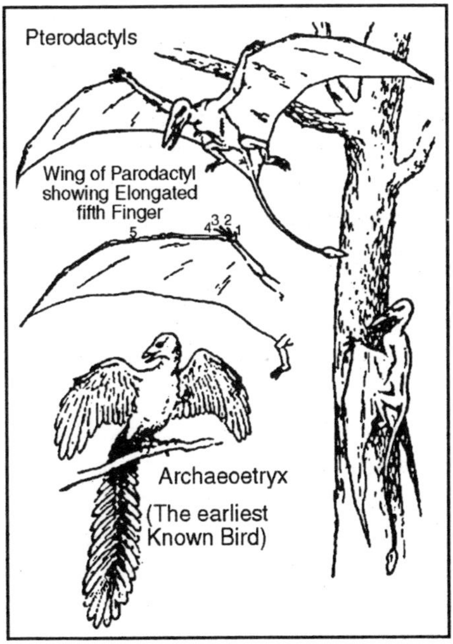

Fig. Flying reptiles and the Brimitive Bird Archaeopteryx

The Pterodactyls had membranous wings like bats, supported both by fore and hind limbs, but mainly by the 'little' finger. The remaining fingers could still be used as claws. Archaeopteryx had true feathers, and wings supported only by the fore-limb, but it retained teeth, a long tailskeleton feathered on either side instead of the tail-fan of modern birds, and three clawed fingers.

Birds are on the average notably smaller than mammals. This is due to purely mechanical aeronautical limitations too complicated to discuss here; as a matter of hard fact, the largest birds capable of flight—swans, vultures, or albatrosses—weigh well under 50 kilograms, while the weight of the great majority is to be reckoned in ounces or even grams, the smallest humming-birds weighing a little under 2 grams.

It is interesting to compare the success of reptiles, birds, and mammals in different zones of the earth's surface. Reptiles have the temperature of their surroundings; consequently their activity is somewhat more than doubled for each rise of 10°C. In the arctic they could scarcely ever be active at all, but would have to exist in a state of almost continuous hibernation, and there are in point of fact no reptiles in the arctic. In the temperate zone they must waste half their life hibernating, and even in the summer cannot compete in activity with a warm-blooded creature; so here reptiles are few and small. In the tropics, however, their average speed of living is more nearly that of a mammal, and they can be active all the year round; in the tropics, therefore, reptiles are more abundant and of greater size—crocodiles, iguanas and other large lizards, giant turtles and tortoises, boa-constrictors and pythons, are all tropical.

Mammals, owing to merely mechanical reasons, can attain to larger sizes than birds. On the other hand, they cannot move readily from one zone to another. So it comes about that in the arctic the birds are the dominant vertebrates, because they can leave in the winter; arctic mammals are few, and almost all of them, like seal, walrus, polar bear, and whale, are entirely or chiefly aquatic. The mammals on the other hand are the dominant group in temperate and subtropical regions.

There are one or two points in connexion with the evolution of mammals that are worth mentioning here. What feathers have been to birds, hair has in part been to mammals; hair and milk together are the mammalian characteristics *par excellence,* hair permitting a constant temperature, milk implying a long period of care of the young after birth. Hair and milk have given mammals the victory over reptiles. But within the mammalian stock itself progress has depended chiefly on two other factors—brain and prenatal care.

Fig. A Female Duck-bill Platypus (Ornithorhynchus) Suckling Its Young

There are three grades of pre-natal care to be found in the single class of mammals. The duck-bill platypus and echidna lay eggs like any reptile. The Marsupials, such as the kangaroo and opossum, nourish their young within their uterus; but the mechanism is not elaborate enough to permit of its being effective after the embryo has grown to a comparatively small size.

To meet this difficulty, the pouch has been evolved. The embryo is born very small and very unformed (a new-born kangaroo is less than two inches long, naked and blind, the limbs not yet provided with fingers).

It can, however, crawl into the pouch; there it becomes glued to the nipple until it reaches a stage more or less similar to that at which higher mammals are born, then becoming detached but still spending its time in the pouch.

Note the enormous enlargement of the jaws (*pmx, mx,* and *md*) for the attachment of the straining apparatus of 'whalebone'; the fore-limb converted into a paddle (*sc,* shoulder-blade; *h,* humerus; *r,* radius; *u,* ulna); the vestigial hind-limb and girdle embedded in the flank (*p*, pelvis; *f*, femur; the other parts of its skeleton have disappeared); the reduction of the neck (the seven neck-vertebrae, *cu*, are fused into one bone); the fish-like shape, and the tail placed horizontally for rapid diving. The typical or Placental mammals, on the other hand, while still retaining the milk diet for their young after birth, all possess the wonderful arrangement known as the placenta, by means of which a huge network of blood-vessels formed by the embryo interlocks in the wall of the uterus with a similar network formed by the mother. By this means, although there is no actual passage of blood from mother to embryo, perfect interchange and nutrition is provided, and the embryo can be protected until fully formed.

In some whales the young are retained within the mother's body till they are over twenty feet long. Whales, being aquatic, can attain to much greater size than any terrestrial animal. They include by far the largest animals which have ever existed, at least twice the bulk of the largest extinct reptiles. They also show interesting traces of their origin from land forms in vestigial hind-limbs.

The Placentals have become as dominant over the Marsupials, wherever the two groups have come into contact, as the Mammals over the Reptiles. Only in Australia, which was cut off from the rest of the world by some earth-movement after it had received an invasion of Marsupials but before it had been reached by any Placentals, are the Marsupials dominant—because without placental competitors.

It is very interesting to find that the Australian Marsupials have evolved into a great many forms not found elsewhere, whether fossil or alive, and that the types evolved are often superficially very similar to those of Placental mammals. There is a marsupial wolf, a marsupial mole, the wombat is like a cross between a badger and a bear, some of the phalangers are not unlike squirrels.

The kangaroo, it is true, is of very different construction from any large Placental. It has filled the niche of herbivorous quick-moving animal, but has filled it in a different way from horse or deer.

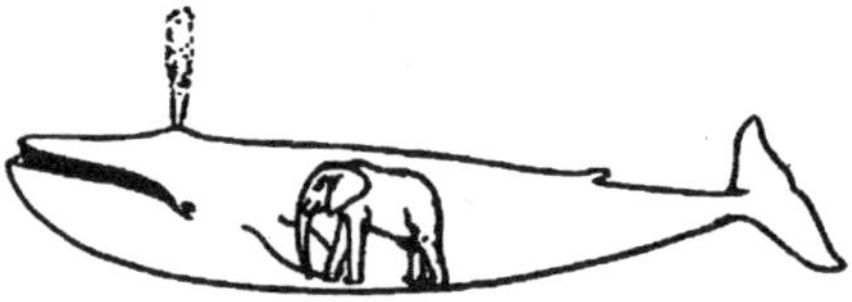

Fig. Sulphur-bottom Whale, *Balaenoptera sulphureus*, 87 feet long, with the African elephant 'Jumbo', *Loxodonta africana*, 11 1/2 feet high, drawn to scale

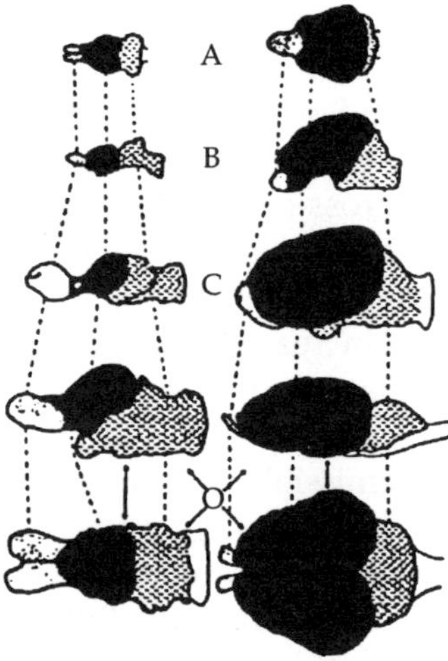

Fig. The Increase in Relative Size Of Brain During The Evolution of the Mammals

The brains on the left (reconstructed from casts of the interior of the skull) are those of mammals from the early Tertiary period. Those on the right are those of living mammals of about the same total bulk. The two brains of each pair are drawn to the same scale.

Arctocyon (a primitive carnivorous form) A Dog

Phenacodus (a primitive ungulate form) B Pig

Coryphodon (an extinct heavy herbivorous type) C Rhinoceros

Uintatherium (related to Coryphodon) D Hippopotamus

The modern brains, in addition to their increase in absolute size, show an alteration of proportions, the olfactory lobes being relatively rather smaller, the cerebral hemispheres relatively much larger. There are, in fact, the same niches to be filled the world over, and different types may fill them in ways superficially alike or superficially different.

Then there is brain. This has played its chief role in the intense competition which took place in the Placentals. Throughout the Tertiary period new lines of evolution were being developed with great rapidity. The upper limit of size was being increased (as, for instance, in the evolution of the horses, the elephants, the whales, the great cats), and physical specialization, especially of teeth and limbs, was being perfected. Both size and physical specialization, however, soon reached a limit. The supporting power of a limbbone is proportional to its cross-section, i. e. to an area; while the weight to be supported varies as the volume. For purely mechanical reasons, therefore, the limb-bones must become relatively larger with increasing absolute size until they finally grow so unwieldy that size no longer pays.

A rhinoceros or an elephant is near the upper margin of size mechanically permissible with advantage to a land animal. As regards specialization, the leg and foot of a horse or a deer, or the teeth of a lion or a cow, could not be much better adapted to their functions than they are now. But, if the instruments at the animal's disposal could not be improved, the methods of using them might be—and this is possible by an improvement in the structure of the brain. The final changes which led to man's evolution seem also to have been primarily brain-changes. Probably, the first divergence of the future human stock from the ordinary land-living mammals came when some shrew-like insect-eating animal took to living in trees.

From some creature like this the Lemur type probably developed, from this again the monkey type. From the old world monkeys, the true apes have clearly descended, by loss of tail and increase of brain-power, and there is no doubt that from some creature which, though not any of the existing apes we know, would have to be classified in the same group with them, man finally evolved. True apes, like the Chimpanzee, are very intelligent and educable.

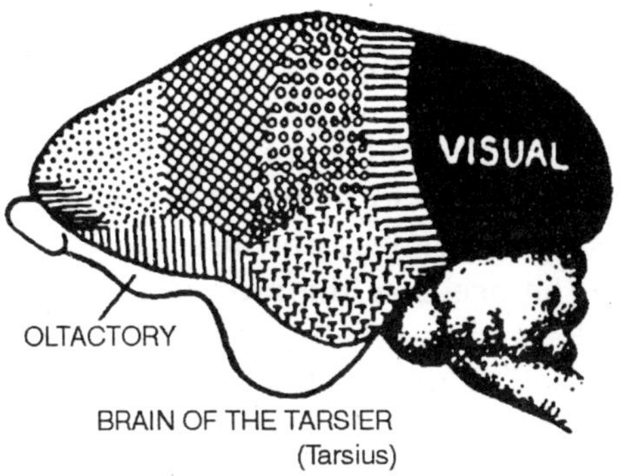

Fig. Development of the Brain and Its Different regions in animals on various evels of development not unlike those which mans ancestry traversed

Above, two Insectivores. Note the much greater development of the visual region and reduction of the olfactory region in the arboreal Tree-shrew as compared with the terrestrial Jumping Shrew. Centre, the Tarsier, the Lemur nearest to the monkeys; below, one of the most primitive true monkeys. The same tendencies are continued and accentuated. In addition, note the enlargement of the pre-frontal area serving for association.

Taking to the trees appears to have been the necessary preliminary to this long evolution. This put a premium on accurate vision and movements which had to be complicated and accurate if the creature was not to fall and lose its life, while the ordinary land mammals continued utilizing smell more than sight, and turning their limbs into mere supports and running organs.

Only in the trees will there grow up the practice of handling objects carefully, and checking the results by careful examination with the eyes, and this eventually led to the development of a true hand and to the manual skill of human beings. It is interesting to find that the parts of the brain connected with sight and manual dexterity increase in size as we pass from lemurs up through apes to man, while the centres connected with smell decrease very much in relative importance.

But physical acquisitions react upon the mind. The monkey has power of examining an object accurately by touch and sight. As is always the case, it is pleasant to indulge a power that we possess; and hence, it appears, the development of that extraordinary curiosity we all know in monkeys. Their curiosity is largely aimless and useless, but if it could be harnessed to the needs of the race, it might yield the most valuable results, and as a matter of fact, this curiosity was the necessary basis of all man's philosophy and science.

Man himself in all probability developed in some temperate and comparatively treeless region, where the surroundings forced him down out of the easy retreat afforded by the tree tops, and compelled the development of skill, foresight, and reasoning power to cope with the animals that were his enemies and those which, in the absence of fruits, he would have to use for food. The rest of the ape-stock remained in its tropical forest home and was never forced to develop further. Remains of a real link between apes and men, the Pithecanthropus, have been found in Java. In the earlier period of human existence, several species of man, some definitely more simian than any types known to-day, were evolved. But to-day only one species survives.

Man probably originated in the Pliocene. We may mention that prehistoric man is known chiefly by the stone implements which he has left behind; in these a slow but gradually accelerated progress is found with the passage of time. He had to survive the Glacial period, an unfavourable environment which probably served to sharpen his wits; and only about ten thousand years ago at the utmost did he discover the use of metals or the methods of regular agriculture.

After the development of the cell, and the origin of sex (which made variation easier when needed), the first necessary step was the aggregation of cells to form many-celled organisms; without this, neither convenient size nor sufficient division of labour would have been possible. The next steps were precisely those of increasing total size and increasing division of labour among the organs.

First came the establishment of two and then three main

layers with different functions, and at the same time the increasing importance of the head end, due to bilaterality and the formation of a nervous system with a dominating region or primitive brain in front. The development of bloodsystem and coelom obviated the need for branched organs, and made much greater size possible. Segmentation again increased the possibilities of division of labour.

Increased size made necessary special organs such as heart and gills, while more rapid locomotion was only possible if better sense-organs and better nervous co-ordination was brought about. Ductless glands made possible a new chemical coordination, especially valuable in regulating growth.

Then the emergence from water to land provided a new freedom; temperature-regulation made life stable, and was an absolute necessity for any delicately adjusted mental life. The development of a complicated brain with emotional moods controlling action, made courtship necessary between the sexes; and out of this has developed much of our sense of beauty. The need to develop out of water produced the reptilian egg; and the further prevention of waste of infant life was brought about by the internal development of mammals, permitting the young organism to come into the world at a greater size. As the mechanical efficiency of the organs of the body approached perfection, an increasing premium was put upon more efficient ways of using the organs—in other words upon brain-power.

The most important development in this respect was improved power of learning by experience. But to learn by experience, the youth of the species must be protected and sheltered; hence the extension of parental care to the young for ever longer periods after birth, and the co-operation of both male and female parent in these duties.

Out of this sprang the family, and the constant association on a common task doubtless made the need for communication more urgent, and so was a necessary step towards speech. Then came arboreal life, and the development of dexterity of movement, of the examination of objects by touch and sight, and so of curiosity. Then the re-descent to the ground, with necessity for great self-reliance and skill, and the harnessing of curiosity to be the basis of organized knowledge; with necessity too for more co-operation, and hence of speech, through which alone organized society became possible.

This brief sketch will perhaps give some idea of the strange series of processes, many of them apparently unconnected, which have yet been necessary for human beings to arise, and for mental activity to become the controlling factor in evolution.

In conclusion, it should never be forgotten that man is, biologically speaking, quite young. The half-million or million years for which he has been in existence constitute but a small fraction of the time which the non-human mammals, for example, took to reach their highest perfection. Nor is there any reason whatever to suppose that his evolution is over.

However, the one great difference between man and all other animals is that for them evolution must always be a blind force, of which they are quite unconscious; whereas man has, in some measure at least, the possibility of consciously controlling his evolution according to his wishes.

But that is where history, social science, and eugenics begin, and where zoology must leave off.

Chapter 13

Fish Farming

Fish culture is the rational cultivation of fish in a confined water area where the practices of both agriculture and animal husbandry are applicable. The soil and water management aspect of fish culture practice involving application of organic manures and inorganic fertilizers for the production of microscopic plants, the phytoplankton, is basically similar to agriculture while husbandry of fish such as feeding, breeding and health care is more or less similar to a livestock farming system.

This farming system is also unique in that the farmed animal is cold-blooded or poikilothermic and lives in a water medium. Although this fish farming is approximately 2000 years old, the importance of it has been realized only recently in the face of mounting pressure on land resources and scarcity of animal protein for the ever increasing human population.

FISH AS FOOD

Malnutrition and starvation are the two serious problems being faced by millions of rural poor in most of the developing countries. The problem of malnutrition is in fact more serious and of a bigger dimension than the starvation problem and is caused mainly due to animal protein-deficient diets. Animal protein is essential for proper growth, repair and maintenance of body organs and tissues.Fish contain about 16–20 per cent protein compared to about 12 per cent in egg, 3.5 per cent in milk and 6–8 per cent in rice and wheat. Moreover, it is wholesome, tasty, highly nutritive and an excellent source of essential minerals, vitamins and essential amino acids.

At present about 31 per cent of the total animal protein supply in the Asian region is in the form of fish protein. For the poorest segments of the population, fish is not only the most important animal protein source, but often the only one.

MINIMAL WATER REQUIREMENT

The reproductive potential of fish compared to any other farmed animal is also very high. A kilogram of female cultivable carp species yields on an average about 0.1 million eggs, each of which has the potential to become 1 kg fish in about a year.

No livestock animal possesses this magnitude of fecundity. Although fish needs water as a medium to survive and grow, it consumes minimal quantity of water compared with any livestock or agricultural crop. Fish also enriches the water with its voided metabolites thus making the water more productive for agriculture.

Low Energy Requirement for Protein

Fish culture systems require a relatively less amount of energy for protein production than any other farming system. Carp culture, depending upon culture practices, requires energy at the rate of 22–468 KJ/g of protein production while a land animal farming system needs over 550 to 3 400 KJ/g.Fish are cold blooded or poikilothermic animals. In other words they cannot maintain a constant and high body temperature like other livestock animals.

Instead, their body temperature fluctuates according to the surrounding temperature. In warmer climates, their metabolism accelerates and they grow faster, while in colder climates, the metabolic rate slows down, resulting in a reduced rate of growth. In this way they save energy by not spending it for maintaining a higher and constant body temperature.

AQUACULTURE PRODUCTION POTENTIAL

Although the world's total marine production now stands at more than 80 million tons per year, all trends indicate that a saturation point is quickly approaching. During the past decade the growth rate hovered around 2 per cent, much lower than earlier decades. On the other hand, tremendous potential exists in aquaculture.

Aquaculture presently produces over 8 million tonnes of fish and shellfish annually.It is estimated that Asian aquaculture production could be raised to 20–30 million tonnes a year by the end of the century. Aquaculture production has increased at an annual growth rate of nearly 7 per cent between 1975–84.

PRINCIPLES OF FRESHWATER FISH CULTURE

POND ECOSYSTEM

Like agriculture, fish culture is also based on a series of processes involving reception and transformation of solar energy. In the pond ecosystem solar energy is utilized for primary production by chlorophyll-bearing plants such as planktonic algae and macrophytes. This conversion of solar energy into chemical energy (food) is guided by the photosynthetic and chemosynthetic activities going on in the aquatic plant community and the rate at which this is carried out is called primary productivity of that ecosystem.A part of the primary production is cycled through different trophic levels resulting in fish production.

Here comes the community of consumers that comprise microscopic as well as large animals, which are unable to synthesize their own food and feed upon primary producers. Different forms of pond life are linked together through predator-prey relationship.

This chain of food production, which follows a general pattern, primary producers-herbivores-carnivores-appears too simple and straight. But, in fact, it is a complex food web with various cross linkages.Fish populations may be classified into several trophic levels, depending upon their position in this food chain. Phytophagous fish such as grass carp and silver carp belong to the second trophic level as they feed upon the first trophic level organisms. Likewise, zooplankton, feeding upon phytoplankton, also belong to the same category.

Carnivorous fish communities thriving upon zooplankton or herbivorous fishes occupy the third trophic level while other predatory fishes preying upon carnivorous fishes belong to the fourth trophic level A relatively simple food chain operates in fish ponds, but a complex one occurs in lakes and other larger aquatic ecosystems.

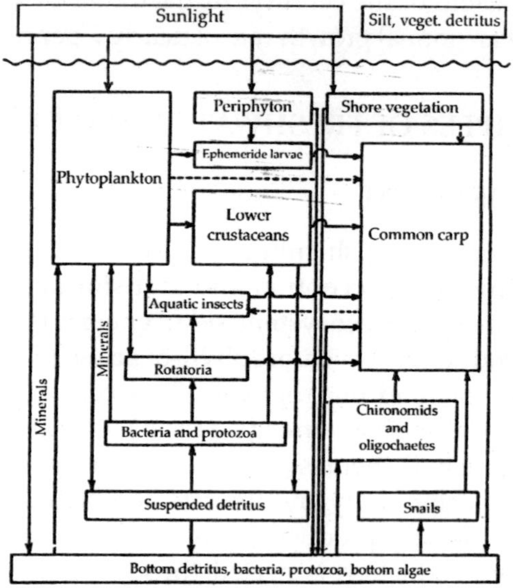

Fig. Pond Ecosystem

The picture becomes even more complicated in large water bodies such as rivers and seas where complex food chains are referred to as food webs which in fact represent several interconnected food chains. There are some fishes which occupy mixed positions, between different trophic levels. They consume both plants and animals and as such, cannot be naturally categorised into any one particular trophic level.

A properly managed pond presents an example of a simple food chain under simple conditions. Here the number of food chains is reduced by encouraging the growth of phytoplankton. The macrophytes such as rooted green plants, floating plants, etc., are not allowed to grow. Phytoplankton is consumed by the zooplankton in the water column, whereas its detritus is utilized by benthic invertebrates. Phytoplankton, zooplankton, detritus and benthic organisms serve as food for the stocked fishes such as the desired carp species. Thus, as much of the available solar energy as possible is utilized for fish production by proper pond management.Primary productivity is dependent on light, carbon dioxide, temperature and essential nutrients, each of which can be a limiting factor.

Of these factors affecting primary production in ponds, the one that can be manipulated easily is the quantity of nutrient elements through the application of nitrogenous, phosphatic and potassic fertilizers, as in agriculture. In ponds, only the top 2 to 5 cm of soil is concerned with nutrientionexchange and the soil below plays a negligible role in the production cycle.Undrainable ponds receive dissolved nutrients and sedimentary particles carried by rain water from the catchment area.

Besides, production and decomposition of minute plant and animal organisms in ponds also modify the properties of the pond bottom to a great extent. The nature and quantity of fertilizers determines the species composition to be used in a culture system. At low phosphate concentration, diatoms are common, but with increasing concentrations green algae become more frequent, eventually giving way to blue-green algae. In addition, excessive phosphate gives rise to phytoplankton blooms which check the light penetration and thus lower the pond productivity through 'autoshading'.

Light energy is one of the major inputs in primary production and hence the success of fish culture depends largely on the efficient utilization of incident light. When incident light strikes the water surface, it is partially reflected and partially transmitted into the water where part of it is utilized in the process of photosynthesis and the rest is scattered or absorbed by suspended particles. In turbid waters, more light is scattered or absorbed, thus allowing the light penetration only to shallow depths.

The rapid disappearance of light in such waters affects

adversely the growth of diatoms. The bottom layer of water, being devoid of photosynthetic plants and also being in close contact with the decaying organic matter, suffers from oxygen depletion causing critical stress conditions for the fish.

Thus, it is important that primary producers must provide oxygen to support the total biological respiration during darkness and also during the less favourable (warmer, overcast or rainy) days apart from providing food for the second and third trophic-level fish.

Table. Solubility of oxygen under different temperatures at 760 mm of Hg pressure

Temperature (°C)	*Solubility of oxygen (mg/l)*	*Temperature (°C)*	*Solubility of oxygen (mg/l)*
15	9.76	26	7.99
16	9.56	27	7.86
17	9.37	28	7.75
18	9.18	29	7.64
19	9.01	30	7.53
20	8.84	31	7.42
21	8.68	32	7.32
22	8.53	33	7.22
23	8.38	34	7.13
24	8.25	35	7.04
25	8.11		

The concentration of dissolved oxygen in the water, which depends on the temperature, is an essential component of the aquatic environment to govern the carrying capacity of a pond. Variations in concentration of dissolved oxygen may occur due to the following three important factors:

- The saturation level of oxygen in water decreases as the temperature rises;
- Supersaturation is an unstable state and
- Plants not only photosynthesize to produce oxygen, they also respire and consume oxygen.

The saturation value for dissolved oxygen available for fish life at 20°C water temperature is more than that at 30°C at a particular atmospheric pressure Dissolved oxygen (Do) concentration is always high at lower temperatures and gradually decreases with increase in temperature.In natural waters, including undrainable fish ponds, DO values are constantly changing because of biological, physical and chemical processes.

The air above the pond water surface may be considered to have a more or less constant percentage of oxygen. However, the partial pressure of oxygen in the air may vary slightly at a given location because of differences in atmospheric pressure. Transfer of oxygen from air to water will occur when water is undersaturated with DO and oxygen will diffuse from water to air when water is supersaturated with oxygen. However, the diffusion of oxygen into the pond water is very slow, except under conditions of strong turbulence, hence the most important source of oxygen is that generated during photosynthesis.Sometimes, phytoplankton blooms or algal scums limit light penetration causing reduction in photosynthetic rates, even in waters with adequate nutrient concentrations.

Oxygen production by phytoplankton is greatest near the surface and decreases with the increase in depth because of self-shading. When heavy infestation of aquatic weeds and dense bloom of plankton occur, the situation becomes much more complex. On the other hand, these are additional sources of oxygen at daytime; but on the other hand, they also respire and consume oxygen throughout day and night. At times the pondwater is supersaturated with oxygen during the day, which is a highly unstable state, while during the night, a greater proportion of oxygen is used up for their respiration, thereby reducing the availability of oxygen to fish. Thus, it creates a wide fluctuation in the level of dissolved oxygen, adversely affecting fish life.

In fact, under such situations oxygen production increases to its maximum during the daytime leaving surplus for the fish even after consuming for their own respiration, but at night this surplus level drops down to critical level.Under conditions of heavy algal blooms and weed infestation, the phytoplankton and aquatic weeds actually consume more available oxygen during day and night than they produce during the whole day During cloudy days, when the incident light is inadequate for phytosynthesis, the situation in terms of availability of DO becomes worse.Aerobic decomposition of organic matter by bacteria is also an important drain on the oxygen supply in ponds.

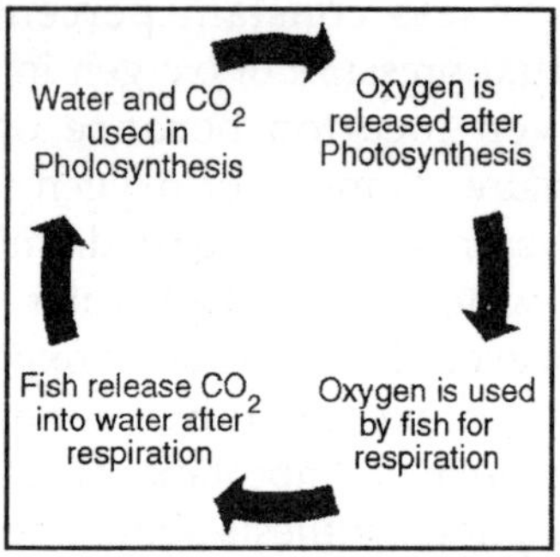

Fig. Oxygen Cycle in Pond

Aerobic decomposition requires a continuous supply of oxygen and proceeds more rapidly when DO concentrations are near saturation. However, decomposition also occurs under anaerobic conditions, but the rate of degradation of organic matter is not as rapid and complete as under aerobic conditions. Under aerobic condition, the end product of decomposition is primarily carbon dioxide.

At times high rate of bacterial decomposition of dead organisms and other organic bottom deposits lead to a condition favouring the increase of the level of carbon dioxide and other abnoxious gases, with a simultaneous depletion of DO, resulting in fish kills and planktonic collapses Therefore, it is important that the pond water should provide adequate oxygen to support the total biological respiration during the hours of darkness.

EMPLOYMENT POTENTIAL

Aquaculture is also considered to be a potential source of employment for poor farmers and displaced capture fishermen. Rapid development of aquaculture has already generated considerable employment through culture of marketable fish, fish seed production and marketing of fish and fish seed.The National Agriculture Commission of India while estimating the employment potential of fish culture has indicated that every tonne of fish produced provides employment to 2.5 persons.

The other important advantages of fish farming are that the production is carried out within easy reach of consumers and also the harvesting can be adjusted to demand, thus minimising distribution problems and spoilage.Rural ponds in Asia, hitherto producing at subsistence level, have succeeded in increasing production per unit area through improved culture practices involving higher stocking densities, polyculture combinations, pond manuring and feeding.

The switch over from monoculture practice to polyculture has significantly contributed toward higher production and the prospect of polyculture appears very bright as the fish seed of desired species is becoming easily available due to the establishment of a large number of hatcheries.During recent years, advances have also been made in traditional aquaculture systems practised in rural India by the development of composite fish culture, a system of polyculture of a group of complementary and supplementary freshwater species of fast growing carps in undrainable ponds.

With the successful demonstration in different agroclimatic zones of India, gradual improvements in technology have been made and it is now possible to obtain a production rate of over 10 t/ha/yr in experimental ponds and up to about 5 t/ha/yr in farmers' ponds against the traditional average rate of production of 600 kg/ha/yr.

To meet the increasing demand of seed of culturable carps, hypophysation techniques have also been developed for both Indian and Chinese carps and as a result they are now being bred in captivity even by fish farmers in remote villages. The emergence of this culture technology suitable for undrainable ponds and the simultaneous development of hypophysation techniques for fish seed production has completely revolutionized fish farm productivity. This manual intends to provide the basic concept and practical guidelines of fish culture in undrainable ponds. Since it is prepared especially for extension agents and field workers, certain important things have been repeated and at times experimental results have been simplified with a view to making it more practical, simple and illustrative.

It outlines the practices of procurement and propagation of fish seed, rearing of spawn to fry and fingerling stage and production of table-size fish following simple sequential steps. Like other farming systems this culture system is also prone to certain unexpected hazards for which one has to be prepared and properly equipped.Such hazards are disease outbreaks, oxygen depletion, pollution, flood, drought, poaching, etc.

The content of this manual is a synthesis of the author's personal field experience, the information gathered from published literature and the observations of other workers in India.Based upon Indian experience, water resources in other developing countries with similar agroclimatic conditions may be utilized for the development of fish culture.It is hoped that this manual will serve as a practical guide to extension workers in popularizing freshwater fish culture in undrainable ponds.

FISH SPECIES FOR CULTURE

The choice of fish species is very important in maximizing production, both in terms of quantity and quality.Since considerable amount of energy is lost in successive trophic levels of the food chain, efficient fish culture always aims at making the chain as short as possible. Because of this, herbivorous fishes are always preferred to carnivorous fishes, the latter being mostly excluded because of their longer food chains. Mixed species farming or polyculture yields a higher production than single species farming.

It is obvious that any single species cannot utilize all the available food in a pond because of its specific feeding habit and hence a combination of compatible species with complementary feeding habits are usually stocked to make better use of the natural food available in the pond. Selection of the species should be based on the productivity of a pond, availability of artificial food resources, availability of seed and the marketing prospects. The principal considerations in species combination are that they have complementary feeding habits, they occupy different ecological niches, they attain marketable size at more or less the same time, they tolerate each other and they be non-predatory in nature.

A combination of plankton and macrophyte feeders is most usual. Ungrazed phytoplankton is fed upon by zooplankton and to utilize them, the zooplankton feeders are included in the combination. The combination of the phytoplankton-feeding silver carp (Hypophthalmichthys molitrix), the zooplankton-feeding bighead carp (Aristichthys nobilis) and the weed-eating grass carp (Ctenopharyngodon idella) is well known in China and Southeast Asia.

In India, under composite fish culture, six species of fish viz. catla (Catla catla), rohu (Labeo rohita) and mrigal (Cirrhinus mrigala) along with three Chinese carps such as grass carp, silver carp and common carp (Cyprinus carpio) are stocked together so as to utilize most of the fish food organisms present in the pond Other similar combinations may work just as well, but the most important aspect is to try to establish a balance between the species based on the food spectrum of the pond Normally fish production increases with the increase in the number of fish stocked per unit area upto a level and then starts decreasing.

Higher stocking density results in increased total production, as there is better utilization of the available food, but in such cases the individual weight and size is reduced. On the other hand, lower stocking density yields larger individual fish.Proper stocking rate for a pond is that optimum level which results in a given time, usually a year, in a production which is highest in quantity and quality of fish and most profitable. In ponds where no artificial feed is used, the total crop becomes dependent on the primary production and in such cases simply by increasing the stocking density, the increase in the total production is not possible.

Even with supplementary feeding the scope of increasing

stocking density and fish yield is limited; it increases to an optimum level and then starts decreasing.Under crowded conditions fish compete for food and space and are stressed due to aggressive interaction. Fish under stress exhibit decreased feed consumption and slow growth and are predisposed to many parasitic and microbial infections. Increase in stocking density simultaneously increases the total oxygen demand with obvious dangers. In undrainable ponds, accumulation of excretory products of the fish population also suppresses their growth rate. With efficient removal of such metabolites by aerating the pond water, the stocking rate can be increased further, thereby enhancing production.

POND FERTILITY

Organic matter and mineral constituents of the pond soil supply the required nutrients for chemical and biochemical production processes. The pond bottom also provides a suitable environment for the decomposers like bacteria and fungi to mineralise organic components of the pond sediment and release soluble nutrients. Sometimes such nutrients are not available in sufficient quantity in the pond and hence they are added from outside in the form of fertilizers.

Since plankton production is often limited by inadequate quantity of phosphorus which is essential for the assimilation of nitrogen into cellular matter, phosphatic fertilizers are widely used in fish culture. Unlike phosphorus, availability of nitrogen does not depend on the inherent status of soil, since it is brought to the soil by different processes. Nitrogen fixation by azotobacteria, blue-green algae, atmospheric electric discharges and photochemical fixation are some of the potential sources of pond nitrogen. In a tropical climate, the fixation of atmospheric nitrogen by blue-green algae is of considerable importance.

However, when nitrogen is added from outside, its form, ammoniacal or nitrate is also very important. It is advisable to use the ammoniacal form of nitrogen in acid and neutral soils and the nitrate form in alkaline soil.Though there is a considerable loss of nitrogen from the ammoniacal form in alkaline soil, the use of ammonium sulphate in low doses is usually recommended, keeping in view the role of sulphate in reducing the soil alkalinity. Potassium is the other essential nutrient for plant growth. In ponds it is easily available both in soil and water and does not form insoluble salts and is rarely deficient except in acid peaty soil. Yet, a little potassium when added to the pond, stimulates the production of plankton.Organic matter of the pond sediment is also an essential factor regulating the bacterial activity. In this context, the ratio of organic carbon to total nitrogen (C/N ratio) is important. Periodic application of organic manures ensures to a certain extent replenishment of nutrients and also provides an energy base for bacterial activities. Apart from this, the organic matter and the bacterial flora are also directly consumed by zooplankton and some fish species.

DISEASES AND THEIR CONTROL

Various intensification approaches such as increased stocking rates, increased feeding, fertilization programmes, etc., sometimes result in nutrient accumulation, frequent appearance of algal blooms, dissolved oxygen deficiency and other water quality problems in undrainable ponds. As a result of such water quality and environmental problems, the infectious diseases and their control assume importance.A fish farming system is unique in that the farmed animal is poikilothermic and lives in water where respiratory oxygen level compared to air is limited and becomes critical at times.

Further, metabolic waste products, left-out feed materials and organic load of the pond bottom can affect certain exposed vital organs and tissues of fish. All such factors affect fish health and contribute to the risk of disease outbreaks.The above basic facts need careful consideration while planning for freshwater pond fish culture. The habitat of an undrainable pond is very varied and dynamic, but can be monitored and managed for increasing fish production.

LIVING SPACE

It has been observed that under identical conditions of

management levels and stocking density fish grow bigger in larger ponds. In Malaysia, grass carp, Puntius sp. and monosex Tilapia mossambica grew bigger in ponds of a larger area indicating the living space phenomenon In other words, the rate of production in a 0.2 ha pond will be more than double that of a 0.1 ha pond, despite the fact that the stocking rate per unit area is the same and all other management components including the genotype of the stocking materials and ecological conditions remain the same.

The ponds having larger surface area are subjected more often to wind action resulting in greater rate of diffusion of atmospheric oxygen into the water.Larger ponds have other advantages also, viz. better cooling action by wind. In smaller ponds, water tends to stagnate and in hot weather tends to heat up quickly. Though it is preferable to have ponds of a large size, there is a physical limitation. Large ponds are difficult to fill and even more difficult to harvest. There must be an optimum size and shape of the pond to balance size with practicability of management, i.e. large enough to allow proper growth of fish, but at the same time small enough to be manageable.

Supplementary Feeding

With the increase in carrying capacity of the pond either by aeration or circulation of water, fish growth can be increased further by supplementing the natural food with some artificial feed. This is the single most important management component for increasing production. In intensive and semi-intensive culture of fishes, supplementary feeding is indispensable. The quantity of feed and the form in which it is offered affect the rate of consumption. Temperature, dissolved oxygen level, crowding and health condition, etc., affect the rate of food consumption.

CHARACTERISTICS OF DRAINABLE PONDS

In India ponds are relatively small and shallow bodies of impounded water with limited wind action. They may be called perennial if they retain water the year round or temporary/seasonal, if they do so seasonally. They may be further classified as drainable ponds and undrainable ponds, depending upon the drainage facility by gravity.

Drainability imparts a very desirable feature to a pond and some authors prefer to call only the drainable type of ponds as fish ponds. However, Indian experience has shown that experimental fish production to the tune of over 10 tonnes/ha/yr can be achieved even from such undrainable ponds through a proper understanding of the biotic and abiotic components of the ecosystem and adoption of suitable culture technologies. Hence, before adopting any culture technology, it is imperative to have an idea about the basic biology of the pond types in terms of environmental factors, community structure and community metabolism.

UNDRAINABLE PONDS

The periods of 'plenty rain' and 'no rain' usually prevail in regions having undrainable ponds. With the onset of monsoon, torrential downpours sweep across the land and the amount and frequency of rain decrease towards the end of the monsoon. Severe floods may occur, whereas a late monsoon or early monsoon of short duration may result in serious drought. Both flood/ rain and drought influence the ecosystem of the ponds on such lands. Small, shallow and seasonal ponds get filled or dry, whereas deeper perennial ponds exhibit considerable fluctuations in water levels accordingly.

Though these ponds are basically constructed for storing water in such areas for multiple uses, ranging from supplying drinking water for human population, live stock, etc., to supplying water for agriculture, recent trend is to utilize them for fish culture. The description of the undrainable ponds is based upon the studies conducted at the Central Institute of Freshwater Aquaculture, Dhauli, Bhubaneswar, India, under an extensive environmental monitoring programme of rural undrainable ponds.

General Morphometry

Undrainable ponds in general are relatively small,

perennial or seasonal water bodies constructed or excavated for multiple uses. Some of these ponds have proper embankments. They greatly vary in their dimensions ranging from 0.02 ha to over 2.5 ha in water surface area and 50 cm to 250 cm in depth.

Larger ponds are relatively deeper while smaller and seasonal ponds are shallower.Unlike shallow seasonal ponds, the bottom of the perennial ponds is never exposed to sunlight and therefore the whole ecosystem of such ponds is quite different from those of shallow and seasonal ponds. Use of these ponds by villagers for multipurpose provides the source of organic enrichment. Usually the only source of water for these undrainable ponds is the heavy rainfall during the monsoons.

However, in some cases, the pond bottom is cut below the water table so that ground water enters the ponds. As soon as the monsoon ceases, the water level starts decreasing gradually and shortage of water is quite common during the pre-monsoon season. Water is lost from the pond through evaporation, seepage and transpiration by aquatic macrophytes and the trees and shrubs planted along the pond sides. Macrophytes tend to appear in both perennial and seasonal types of ponds but with increased intensity in shallower ponds.Presence of thick sediment layers in the bottom is the most characteristic feature of these ponds which gradually get accumulated during the course of time and vary between a few centimeters to over a meter and half in thickness.

The quality and quantity of sediment deposition depend mainly upon the original soil, method of construction, nature of embankments, macrophyte cover, pond productivity, organic and inorganic additions, species cultured, etc. Although sedimentation is relatively faster in smaller ponds, there is a positive correlation between the age of a pond and its sediment thickness.

Physico-chemical Environment

The water depth and total volume of water available for

individual fish are crucial in fish culture systems. Adequate water depth is needed not only for optimum growth, but also to provide enough space and oxygen for fish life. Water levels in these ponds are mainly dependent upon monsoon rains.After the monsoon season, the water level starts decreasing gradually and shortage of water is quite common during the summer season which is the most crucial time for fish culture since the fish growth rate is faster in this period. In fact, during the time of lowest water level the ponds contain the maximum biomass.

In shallow and seasonal ponds, sufficient phytoplankton population fails to appear and the soft sediment layer is vigorously stirred up by fish, making the water more turbid, thereby reducing the photosynthetic process by limiting light penetration. Eventually, the total amount of available dissolved oxygen may not be, at times, sufficient to meet the demand for total community respiration and the chemical oxygen demand of the sediment, resulting sometimes in mass fish kill and planktonic collapse.

On the other hand, in deeper perennial ponds where the water column is more than 3 m, fish life is again adversely affected. In such ponds the photosynthetic or oxygen producing zone is less in comparison with the oxygen consuming layer.In addition, the sediment proper and the sediment community also consume a considerable amount of oxygen. All such conditions lead to a negative oxygen balance. The water in most of these ponds remains slightly alkaline (pH 7.0–9.0).

The NH_4-N (ammonia-nitrogen) content of the waters remain below 0.02 mg/l with even lesser quantities of No_3-N (nitrate-nitrogen). The PO_4^-P (phosphate) concentrations remain low and these chemical features of such ponds suggest that these waters are highly nutrient-deficient, particularly in nitrogen. On the contrary, the pond sediment is rich in organic and inorganic nutrients.The organic carbon ranges between 3 (in newly excavated ponds) and 50 mg/g dry sediment weight (in older ponds).

The nutrient status of the sediment differs completely

from that found in the overlaying water column In general, all the basic nutrients in the pond sediment are about thousand times higher than in their respective water column.

Carbon dioxide and oxygen are the most important gases affecting the pond community including fish. During the photosynthetic activity, carbon dioxide is usually at zero level while during the darker period its concentration increases. At higher concentrations it may be toxic to fish life. Carbon dioxide toxicity increases with decreasing level of dissolved oxygen. Carbon dioxide concentration can be tolerated upto 20–30 ppm in these ponds provided oxygen is near saturation.

Community Structure and Function

Bacterioplankton and phytoplankton constitute the basic food for the fine filter feeder fish species and also for the zooplankton which form the main food of the rough filter-feeder species. The bacterioplankton population is always higher in those ponds which are associated with the activities of larger human and livestock populations.Most of the relatively older ponds with frequent appearance of Microcystis bloom have higher levels of bacterioplankton population (3–10 million/ml).

On the contrary, the newly constructed and recently desilted ponds have less dense bacterioplanktonic community, around 1–2 million/ml. Macrophytic infestation also significantly limits the bacterioplankton production.

The planktonic detritus originates mainly from decomposing fragments of the phytoplankton and zooplankton and has generally a concentration range of 2,000 – 20,000 number/l. The main groups of phytoplanktonic population are Myxophyceae, Chlorophyceae, Eugleno-phyceae and Bacillariophyceae; whereas copepods, cladocerans and rotifers constitute the majority of zooplanktonic population.

Some of the very old ponds having excessively thick sediment layer face Microcystis blooming. In such ponds the benthic animal fauna is represented by a very small number. In the majority of the ponds the benthic animal communities are dominated by red chironomids and oligochaetes indicating the general oxygen deficiency in the sediment layer.The bacterial decomposition and nutrient recycling in ponds are greatly influenced by the anaerobic nature of the sediment. At the initial stage of Microcystis bloom in older ponds, the oxygen production has been found to be the highest (1) (over 15 g O_2/m^2), whereas the total community respiration remains considerably low (2) (below 10 g O_2/m^2). However, during the active decomposition stage of Microcystis (plantonic collapse stage) the total oxygen production level goes lower (5–6g O_2/m^2) than that of the community oxygen consumption (4) (6–7 g O_2/m^2).

In older ponds, especially those having thick anaerobic sediment, the biochemical oxygen demand ranged between 70 per cent and 90 per cent of the total oxygen production, ultimately causing anoxic condition leading to fish kills.The majority of rural undrainable ponds are characterized by anaerobic benthic sediments. The dead and decaying organic matter settles down to the pond bottom (sedimentation) where it is subjected to further decomposition and mineralisation.The upper layer of the sediment remains aerobic while the deeper layers are deficient in oxygen and thus anaerobic. These perennial undrainable ponds in tropical monsoon lands with yearround warm water under plenty of light offer an excellent possibility for fish culture.

Most of the species cultured greatly depend upon natural fish food resources and with a limited dependence upon artificial supplementary feed.However, without proper environmental management, the water remains infertile due to the overall nutrient deficiency with a very pronounced nitrogen limitation, although they possess a very high production potential. On the contrary, the pond sediments have extremely high level of organic nutrients in almost locked-up conditions which remain unutilized due to the anaerobic nature of the pond bottom. However, though regular raking up of the pond sediment, either by manual or biological means, the organic nutrients could be released for making the pond water more productive.

Proper management methods can optimise fish production in perennial ponds at most economical rates while seasonal ponds can suitably be utilized for fish seed rearing and also for short-term fish production depending upon the duration of water retention. The general feature of the properly managed and ill-managed undrainable fish p and 8b.

DRAINABLE PONDS

Ponds which can be supplied with water and drained of its water according to the requirements of the fish farming operation are known as drainable ponds. These ponds require suitable ground with proper embankments, inlet and outlet structures and adequate supply of water on regular basis.

CARP CULTURE

The most successful system of pond fish culture is the polyculture of three Indian major carp species-catla, rohu and mrigal along with three Chinese carps viz. silver carp, grass carp and common carp. In India this is commonly known as composite fish culture.

The best results in terms of fish production in this system results not only through a judicious combination of species, but also due to appropriate management techniques including pond fertilization, supplementary feeding and health care.On the basis of growth performance of different species, modifications are often made in stocking density, species ratio, fertilization schedule and supplementary feeding programme in different agroclimatic conditions.

Fig. Carp culture

High rates of fish production to the tune of over 5 500 kg/ ha/6 months, 7 200 kg/ha/8 months and over 10 tonnes/ha/yr have been achieved in composite fish culture trials conducted in different agroclimatic conditions of India.The carp culture system as a whole is operated as a three-tier culture system where the practices are adopted for rearing fish during their different stages till they are harvested. Spawn (post larvae) are reared upto fry (2–3 cm) stage in nursery ponds, fry to fingerlings (8–12 cm) in rearing ponds and finally fingerlings to table-size fish in composite fish culture ponds or stocking ponds. Relatively smaller, seasonal ponds are mainly used for rearing spawn to fry stage and harvested after 2–3 weeks.

Several crops of fry are usually taken during the season. Pond fertilization by cattle manure and feeding with 1:1 mixture of oil cakes and rice bran is the usual practice. Fry raised in nurseries are reared upto fingerlings in slightly bigger ponds (0.05 – 0.1 ha) of seasonal or perennial in nature. Fingerlings are removed after 3 months and stocked in composite fish culture ponds.

INTEGRATED CARP FARMING

An integrated approach of composite fish culture together with compatible combination(s) with poultry, duckery, pig rearing and cattle raising is now being adopted.

Table. Annual production through integrated carp livestock farming

Integration	*Fish production*	*Animal production (live weight)*
Fish + Pig farming	6 – 7 ton/ha	4 000–5 000 kg pig meat
Fish + Duck farming	3 – 4 ton/ha	500 kg duck meat + 17 000–20 000 eggs
Fish + Poultry farming	4–5 ton/ha	60 000–70 000 eggs + 1 500–2 000 kg meat

Under this system of farming small livestock and farm yard animals, viz. pigs, poultry, ducks, etc., are integrated with composite fish culture by siting animal housing units on the pond embankments in such a way that the animal wastes and washings are diverted into fish ponds for recycling. The fish not only utilize spilled animal feed but also directly feed on fresh animal excreta which is partially digested and is rich in nutrients. Surplus excreta supports the rich growth of planktonic fauna. Fertilizers and supplementary feed are not used, resulting in drastic cost reduction.

Integrated Fish-pig Farming

Pigstyes are constructed either on the pond embankment or near the pond to facilitate easy drainage of waste directly into the pond which acts as pond fertilizer and supports dense growth of natural fish food organisms. Besides, fish also feed directly on the pig excreta. No other feed or fertilizer is applied to the pond. A pond is prepared by following the usual pond preparation techniques and stocked with fingerlings of all the six species of carps cultured under composite fish culture at higher of 8 000–9 000/ha with surface, column, bottom feeders and grass carp in the ratio of 40:20:30:10. Marketable size fish are sold by partial harvesting while final harvesting is done only after 12 months of farming.

About 2 months-old weaned piglets are fattened for six months when they attain slaughter size (60–70 kg) and similarly a second crop is raised within the next six months. About 30–40 pigs should be kept for proper fertilization of the pond. Pigs are fed on mash at an average rate of 1 kg/day. Green grasses or animal fodder is also provided. Grass with interlocked soil in root system (sod) are provided once a week to avoid mineral deficiency. Grass carp is fed with aquatic weeds or green animal fodder. Fish yields ranging from 6 000–7 000 kg/ha/yr are generally obtained.

Integrated Fish-duck Farming

This is also an efficient integrated system based on the principle of waste recycling. Pond preparation technique is basically the same. A duck house is normally constructed on the pond embankment or on the pond water on a floating platform When given free range, ducks feed on aquatic organisms such as insect larvae, tadpoles, molluscs, weeds, etc. The duck droppings like pig excreta act as fertilizer.

Ponds are prepared and stocked with fingerlings of all the six carp species at 6 000 ha with surface, column, bottom feeder and grass carp in the ratio of 40:20:30:10. Fingerlings of over 10 cm are preferred for stocking. About 200–400 ducks are sufficient to adequately fertilize a l ha pond.

Normally 2–3 months old ducklings are preferred. Although ducks are able to feed upon natural food from the pond, they are also provided with duck feed at the rate of 100 g/bird/day. Ducks start laying after 5–6 months and continue for 2 years. Fish yields ranging from 3 000–5 000 kg/ha/yr are generally obtained.

Integrated Fish-Poultry Farming

Under this system of integration the poultry birds are raised in cages under a shed normally constructed over the pond embankments or in the vicinity of the pond. The space requirement in such a system of poultry raising is about 1 sq.ft. per bird. The droppings of the birds fall on the floor from where these are collected and applied to the pond. The chicken house can also be built directly over the pond water so that the excreta may fall in the pond water underneath.

Usually, 400–600 chickens/ha of pond water surface are used. No feed or fertilizer is applied in the pond, except aquatic vegetation for the grass carp. Fish production at the rate of 4–5 t/ha is possible using this system. In India, this system of freshwater fish culture has assumed greater significance in view of its potential role in recycling of organic wastes and in integrated rural develop.

AIR-BREATHING FISH CULTURE

Besides freshwater ponds, there are many low-lying areas which become waterlogged during the rainy season. In course of time these areas get infested with dense aquatic vegetation and turn into swamps. Swamps are also formed along the irrigation canals due to profuse seepage. These areas are best suited for culturing airbreathing fishes such as Koi Singhi, Magur and Murrels without getting involved in costly processes of their reclamation essentially needed for carp culture.

There are three levels of culture practices. low cost culture, semi-intensive culture and intensive culture depending on inputs and level of management Fingerlings (6–10 g) and feed are the two material inputs used in the culture system. Fertilizer is not used. However, replenishment of water becomes an essential input in case of intensive system of culture in ponds where very high stocking rate and intensive feeding is practised to obtain very high yields Monoculture as well as polyculture of these fishes are commonly undertaken, yielding production to the tune of 3 000 to 7 000 kg/ha/yr.

Production at the rate of over 3 000 kg/ha/8 months is possible through monoculture of Channa marulius in swampy ponds. In intensive system of monoculture of magur and singhi with frequent change of water, yields of over 15 t/ha/yr have been obtained. Presence of naturally occurring weeds in airbreathing fish culture ponds not only provide protection against poachers, but also encourage the growth of insects which are consumed by the fish.

SEWAGE-FED FISH CULTURE

The wastes, including sewage and waste water produced by the human community hold high potential for fish production. In India itself there are about 150 sewage-fed fish farms covering an area of about 12 000 ha. Very high production in the order of 7–10 tonnes/ha/yr has been obtained from ponds fed with sewage which invariably contains high percentage of N,P,Ca,K,etc.

An average production of about 7 t/ha/yr is easily obtained using a mix culture of 5 carp species.The sewage fed ponds are generally dewatered completely during summer so as to remove all the carnivorous fishes.

The pond is initially fertilized by introducing partially treated sewage effluent upto about 75–80 cm and then clean water is pumped in to raise the pond water level to 1.5m. Within a month the pond stabilises with respect to dissolved oxygen and becomes suitable for stocking with fish seed. During raising of marketable size fish, additional fertilization with sewage effluent is carried out in small doses every month and the pond is netted frequently to help oxygenate the water and in course of netting the marketable size fish are also harvested.

RENOVATION OF EXISTING PONDS

The majority of freshwater fish ponds in the Indian subcontinent are the dugout ponds of an undrainable nature which at times lack proper embankments.During the course of culture operations, such ponds receive huge amounts of feed, fertilizers and manures as critical inputs and sediment particles carried down by rain water from the catchment area.

A portion of the organic production in the pond also undergoes death and decay and gradually adds to the pond bottom sediment. Thus, with the advancement of time, a thick sediment layer is formed reducing the depth of the pond.They are quite rich in organic and inorganic nutrients, but due to slow bacterial action under prevailing anaerobic conditions the nutrients are almost locked up in the sediment and are not available for primary production. Further, the anaerobic decomposition of the organic matter accumulated in the sediment releases harmful gases and depletes dissolved oxygen level in the water.

Thus, it becomes necessary to renovate the existing ponds periodically every 4–6 years by removing sediment from the pond bottom, redressing and repairing the dykes, etc., in order to make the ponds more suitable and to regain their fertility. For this, the following practical measures are recommended.As soon as the water table of the area surrounding the pond goes down the renovation work can be initiated.

Summer is the most suitable period for this Purpose as complete drying of the water body is possible. In this period pond renovation can be carried out efficiently and economically. Removal of slushy silt from the partially dried pond bottom is difficult, laborious and expensive.

DEWEEDING

It has been observed that most of the rural ponds are not properly managed and become weed-infested in course of time. Before dewatering the pond, large floating weeds such as water hyacinth should be eradicated by pulling them out manually or mechanically. Otherwise, collection and removal of such weeds will require more labour and time. Other rooted emergent or submerged weeds can be taken care of only after draining the pond.

DEWATERING AND DRYING

Dewatering of the existing pond is possible either by draining the water after cutting a portion of the embankment or by pumping out. If the water table in the surrounding area is high, there is considerable inflow of water from the pond bottom. This phenomenon of sub-surface secretion is called percolation.In case the rate of percolation is high, several furrows or ditches may be made towards the lowest contour point where a pit may be dug out to drain all the percolated wall.

Periodical pumping of water from the pit facilitates keeping the bed dry.Where the bed is found to be uneven, contouring is necessary to estimate the amount of silt to be removed. It is done by taking the level measurements at certain spots on the pond bed. It will also help in redesigning the pond taking into consideration the highest flood level and maximum rain water level.

DESILTING

After complete dewatering, the pond bed is allowed to dry and develop cracks in the silt mass. The texture of silt is different from that of the bottom hard soil and cracks quickly. At this stage dried silt is cut and removed manually or mechanically and heaped at a suitable place for its utilization in agricultural fields. Where complete drying is not possible due to high rate of percolation, walking platforms made up of bamboos or wooden planks may be put on the slushy bed to facilitate desilting work.In some larger ponds it becomes difficult to dry the central portion of the pond bottom as it is nearer to underground water table. In such cases the slushy and loose silt should be scrapped and spread to the sides with the help of wooden planks tied with ropes for pulling. This helps in drying the silt and easy removal thereafter.

RECLAMATION OF WATER BODIES

Derelict waters in millions of hectares, lying unutilized, are common sights in most of the South Asian countries. Such untapped water bodies with potential for aquacultural production may be reclaimed and made suitable for fish culture by adopting more or less similar procedures.In case of larger water areas, it would be better if they are connected temporarily to nearby natural or man-made drainage systems having relatively lower bed level for complete dewatering by gravity and making the entire area completely dry.

However, if such topographic facilities are not available, heavy duty water pumps may be put into use for quicker dewatering.In extensively large areas dewatering by draining or by pumping is not feasible. Moreover, the dry period of the year also may not last long enough to permit the work to be completed. It has been experienced that such areas can also be successfully reclaimed and renovated by partitioning into smaller units by raising cross bundhs, farm roads, etc. Each newly formed unit then can be dewatered, dried and desilted.

MAINTENANCE OF DYKES

In general, rural ponds lack proper embankments. During high rainfall or peak irrigation periods in canal-irrigated areas such ponds get inundated with water from the neighbouring agricultural fields causing stocked fish to escape, predators and unwanted species to enter and at times results in mass fish kills due to pesticide pollution. Hence, provision of proper dykes is a must.

Existing pond dykes should be repaired every year after the monsoon.Rats and crabs cause great harm to pond dykes by making holes. Such holes allow serious leakage and if not checked immediately, may endanger the stability of the dykes. Periodically and especially at the time of renovation, such spots should be properly repaired by stuffing binding clay, claylime mixture or any other locally cheap cementing material.

Due to poor consolidation, erosion from the top of the dyke during heavy rains usually results in grooving out of small channels. These areas should be covered with earth, levelled, thoroughly rammed and grass turfed. In relatively larger ponds, wave action due to wind also causes large-scale dyke erosion.

By putting large floating aquatic plants such as water hyacinth along the sides of the dykes exposed to wave action during the windy season such erosion can be checked. Frequent erosion in steep dykes during heavy rain or wind can be avoided by strengthening the inner sides of the dykes with poles or bamboos or corrugated cement planks.

Most of the traditional pond dykes are below the required height; as a result, overflow of water occurs during heavy rains or flood. These dykes should be properly raised and the height may be kept at a minimum of one meter above the maximum water level recorded in that area. While raising the dykes, the top width may be kept at a minimum of 1.5 m with 2:1 slope (horizontal: vertical).

Cutting the dyke to allow water into the pond from the surrounding area without any secured screening is a normal practice, which however creates many management problems. The silt mass is very rich in organic and inorganic nutrients making it most suitable for application in agriculture and horticulture. Being non-cohesive and unstable, it is unsuitable for making dykes as it may be washed back in the pond.

CONSTRUCTION OF NEW PONDS AND FARMS

Village ponds, homestead or backyard kitchen ponds, garden or farm ponds, irrigation ponds and occasional ponds such as brick mine pits and quarries, etc., occupy enormous freshwater areas in the tropics and are used for fish culture with minor improvements. However, ponds designed and constructed for fish culture are easier to manage and are expected to give higher production.

Although certain well-defined guidelines do exist for the construction of fish ponds, it is mainly the topography of the site which determines the basic design of the pond/farm. There are, however, certain basic principles to be considered when choosing a site and deciding the method of pond construction.

SITE SELECTION

Selection of suitable sites for fish farm construction is very important. The following three essential conditions guide the proper site selection:

- Topography
- Source of water and its quality
- Soil type

Topography

It is economical and convenient to construct ponds in waterlogged areas, irrigation command areas or in marginal lands. In such areas construction cost is relatively low mainly due to limited earth cutting.For example, a pond of 100 m × 40 m (0.4 ha) of water area requires only 3 234 m^3 of earth to construct around a dyke of 2 m high above ground level (GL) with side slope ratio of 2:1 and top width of 1.5 m.

This quantity of earth may be obtained only from 1.1 m depth of cutting. This limited depth of cutting reduces the construction cost considerably. However, full consideration should also be given to the possible effects of flood.

The surface features of the area proposed for the pond or the farm is also equally important. A saucer-shaped area may be an ideal site for a large dug-out pond, because it may hold appreciable quantity of water with a small amount of earthwork.For smaller and flat areas eye estimation is enough, but for a big area proposed for farm construction with a number of ponds for different purposes and of different sizes, it is essential to conduct contour survey for determining the topography and land configuration. The site should be easily approachable so that there may not be any difficulty in the transportation of input materials and in the marketing of the produce. The labour and materials required for construction and operation should also be locally available as far as possible.From an efficient management point of view the pond site should, if possible, be within the sight of the farmer's house. It also reduces the risk of poaching. Siting fish ponds near the farmer's other agricultural or livestock farming activities makes it easier to integrate all the farming activities.

Dyke

The dyke should be properly designed so that it can hold maximum water in the pond and withstand the hydraulic pressure. The slope of the dyke usually depends on the type of soil. Suitable side slopes for different soil 7.

Table. Suitable Slopes for Different Soils

Soil type	*Soil (horizontal:vertical)*
Clay	1:1 to 2:1
Clay loam	1.5:1 to 2:1
Sandy loam	2:1 to 2.5:1
Sandy	3:1

Provision for a berm of sufficient width may also be provided for stabilizing the slopes. A wider berm also helps in operating the net in the pond. The berm should be 1 m or more in width. The top width of the dyke should be decided taking into account its usage. Usually the minimum top width of the dyke should be 1.5 m.

The wider crest requires not only a larger area for dykes, but also an increased amount of earth material involving heavy expenditure. It is always wise to design the dyke as per the quantity of earth expected to be available from excavation work.

A soil-type containing approximately 25 per cent silt, 35 per cent sand and 40 per cent clay is most suitable for dykes. However, if excavated soil quality is not up to the above standard, provision may be made for a clay core to make the dyke watertight. While designing, about 10–12 per cent allowance may be given for settling of earthwork.

Source of Water

A dependable source of water supply must be available within or near the site, even for undrainable ponds. However, unlike drainable ponds, undrainable ponds require just sufficient water to fill the ponds and to compensate the water loss through seepage and surface evaporation thereafter.

Equally important is the need for avoiding excess water and hence there must be arrangement for the excess water to escape through a bypass channel or a spillway.The water supply to the pond should as far as possible be natural, preferably rain water.

However, alternative arrangements of water supply should be made for dry season either from a deep tube well or irrigation canal or from perennial sources like spring, stream, river, etc. Ponds should be on the lower lands to allow accumulation of surface runoff from a larger catchment area.

However, care should be taken to provide proper bypass or spillway to avoid flooding. A higher subsoil water table due to irrigation in surrounding fields and percolation from artificial or natural channels, in addition to absorption from rain water, also helps in maintaining water level in undrainable ponds.

The quality of the available water is also equally important for fish culture.Pond fish production is influenced by the physical and chemical properties of the water. Water should be clear as far as possible. Turbid waters which carry suspended solids cut the light penetration, thus reducing primary productivity of the pond.

Excess of suspended solids also adhere closely to the gill filaments and cause breathing problems.Water temperature also significantly influences the feeding and growth of fish. Prevailing water temperature, ranging between 15°C and 35°C in tropical areas, is most suitable for carps. The chemical quality of water depends on its content of dissolved salts. Rain water does not carry any dissolved salts. However, it collects nutrient salts from the ground surface of the catchment area. The water should be neither too acid nor too alkaline; neutral or slightly alkaline waters are most suitable for fish culture and hence acid water should be limed to make it neutral.Waters with p^H values below 5.5 or over 8.5 are not proper for fish culture. The farmer will need huge quantity of lime to neutralize it while highly alkaline water may cause the precipitation of both phosphate and iron and if it remains continuously above pH9, it may be harmful to fish.

Soil Type

Pond soil must retain water. Soils with a low infiltration rate are most suitable for fish pond. The filtration rate of different types of soils. The best soils for our purpose are thus the impermeable clay which can be easily compacted and made leak proof.

Table. Infiltration Rates of Different Types of Soil

Soil type	*Infiltration rate (mm/ha)*
Clay	1–5
Clay loam	5–10
Silty loam	10–20
Sandy loam	20–30
Sand	30–100

Loamy soils can also be used, but they need well compacting and may leak slightly in the early stages, although they tend to seal themselves with time. Sandy and gravelly soils should be avoided, but if they are the only ones available they must be made impermeable with a thick coating of clay or with polythene sheeting.

Soil impermeability can also be achieved by soil compaction at the pond bottom and dyke with either a mixture of soil + 1–5 per cent cement or soil + 10–20 per cent cowdung.

Treated areas should be kept moist for 2–3 days by gently sprinkling water to avoid cracking and finally the pond is filled with water. Peat soils have special problems, since they are usually very acidic in nature and need sufficient liming, while the organic matter decomposition may lead to dissolved oxygen deficiency. Soils rich in limestone also create special problems, since the excessive lime content tends to precipitate phosphate and iron.

Such ponds would then have little plankton population and macrophytes and would be relatively sterile. This can be overcome by adding sufficient organic matter such as cowdung, poultry manure, etc.A general and convenient field test for the soil quality is to take a handful of moist soil from the test holes made at the proposed site and to compress it into a firm ball.

If the ball does not crumble after a little handling, it indicates that it contains sufficient clay for the purpose of pond construction. Accurate determination of the composition of the soil and its water-holding character is possible by hydrometer method.Several test holes may be made across the site and soil samples may be collected vertically from every 0.5 m of depth reaching up to a level of 3–4 m in a test hole. Using the results of the soil tests, a soil profile chart for the proposed site may be drawn. An arbitrary soil profile chart is presented showing the presence of clayey soil up to a depth of 3.5 m.

DESIGNING

Based upon the survey on topography, soil type, water supply, etc., the detailed designing and layout of the ponds/ farm are done. However, the following additional points are also to be considered.The production or stocking ponds are stocked with large size fingerlings of about 10–15 cm size in the case of composite fish culture.

To attain this size, the hatchlings are reared in much smaller and shallower ponds called nursery and rearing ponds for about 2–3 months. In the nursery ponds the hatchlings are reared up to fry stage and in the rearing ponds the fry are reared till fingerling stage.

The ratio of water area among nursery, rearing and stocking ponds in a fish farm depend upon the basic objective of the farm. In case of a fish seed farm, only nursery and rearing ponds are to be constructed with a small area for few stocking ponds to be used for raising the brood fish, while in the case of fish production farm only stocking ponds are to be constructed for producing table size fish from fingerlings. There is no hard and fast rule regarding the size of a pond.

However, nursery ponds should be small and shallow. Ponds having 0.02–0.06 ha water area and 1–1.5 m depth are most suitable as nurseries. Rearing ponds are relatively larger, preferably between 0.06 to 0.10 ha in size and 1.5 to 2.0 m in depth. The sizes of stocking ponds vary tremendously. For newly constructed undrainable ponds, total water area of 0.25 to 1.0 ha is recommended.

In shallow ponds the water becomes heated easily. In deeper ponds light cannot reach the bottom. In very deep ponds thermal stratification may occur with colder deoxygenated bottom layer. Dead plankton and faecal matter from fishes may fall on the bottom layer where the nutrients may be locked up.

Table. Practical size and depth of nursery, rearing and stocking ponds

		Depth (m)*	
Pond type	*Size (ha)*	*Irrigated command/ water logged areas*	*Rainfed+ /non -irrigatedareas*
Nursery pond	0.02 – 0.06	1.0 – 1.5	1.5 – 2.0
Rearing pond	0.06 – 0.10	1.5 – 2.0	2.0 – 2.5
Stocking pond	0.25 – 1.0	2.0 – 2.5	2.5 – 3.5

* Excluding the freeboard

\+ May vary depending on impermeable strata at pond bottom

However, in case of rain-fed areas where the water table goes down during the dry season, the depth should be kept around 3.0 – 3.5 m to store more water during the rainy season.Although a square pond is economical to construct for its minimum length of dyke, a rectangular shape of the pond (length:width in proportion of 3:1) is considered to be ideal.

In any case the pond width should not exceed 30 to 40 m as it is difficult to operate a fishing net in broader ponds. The nursery and rearing ponds may be square, since they are too small to pose any problem for netting. The corners must be curved to avoid fish escaping the net during harvesting.

CONSTRUCTION

Before initiating the construction work, proper estimates have to be prepared based upon the design details, which will include the cost of all the materials and the labour. Strict supervision is required at every step of construction to ensure the adherence to specifications laid down in the design.

Time of Construction

If the construction work is taken up at the most appropriate time or season of the year, the work becomes easier and economical. The best time of the year for constructing ponds in clayey soil is post-rainy period and winter when the soil is soft rather than at the end of the dry season when it is very hard.

For swampy and waterlogged areas the most desirable time is the late summer when the area becomes completely dry. However, if a pond is built during winter or early summer and is not filled immediately, weeds may grow and cover the bottom. In such cases deweeding is needed before filling the pond.

Preparation of Site

The site should be thoroughly cleared of all the trees, bushes, etc. Even the roots of trees should be removed. No woody material should be left because the same will eventually rot and cause leaks. Some tree trunks rot very slowly and may cause problems during netting.

Marking the Outlines

This operation involves laying out the features of ponds on the ground in order to mark out the areas from where the earth will have to be cut and removed and also where earth will have to be embanked. Initially, iines are drawn according to the layout, followed by pegging and fixing stakes or posts.

Strings are stretched between the tops of pegs and posts to mark the complete profile of the dyke with its correct height, width and slopes

Pre-excavation Work

Prior to pond excavation and dyke construction, all loose surface soil should be removed from about 20 cm depth within the total outlined area of the dyke and the surface should be roughened by ploughing or digging. In order to unite the body of the dyke to subsoil, it is desirable to dig a small "V" shaped key trench. When the dyke is to be made on a sandy, gravelly or marshy soil base, the construction of a key trench becomes essential and in such cases digging should be done until watertight foundations are reached.

The key trench is a small ditch or furrow dug along the line of the centre of the walls about 0.5 m – 1.0 m wide and 0.5 m deep. This trench is filled in with a good clayey soil and is well rammed. If good clayey soil is not available in the area, ordinary soil should be well compacted into the trench. The purpose of the trench is to stop seepage of water underneath the walls.

Pond Excavation and Construction of Dykes

The excavation work can be carried out within the area marked for the pond bottom either manually or mechanically. However, the final levelling of the pond bottom and sides should be done manually with proper ramming and finishing as per the original design. The construction of the pond becomes economical if earthen dykes are made around the pond using the excavated earth from the pond bed.

All dykes should be raised, dumping the earth layer by layer stretching right across the whole section and in such cases each layer should not exceed 20 cm in thickness. All large clods should be broken and each layer should be thoroughly consolidated by watering and ramming. The sides and top of the dykes should be properly dressed and finished with wooden thappies (wooden block with handle for ramming).

In case the soil quality is not suitable for making dykes, a clay core is provided in the dyke to make it watertight. A mixture of 1:2 of sand and clay is used to make the clay puddle. This should be consolidated, compacted and deposited in 10–15 cm thick layers. Each layer should be adequately moistened before the next layer is laid and precaution should be taken to prevent the puddle from becoming dry and cracking.

Dykes must be well compacted to render them stable and the top should be rammed flat so that small vehicles can also run along when needed. Short creeping grass is recommended to be grown on the top and sides of the dyke. Trees are not desirable since their dense shade inhibits the productivity of the pond.

Water Inlet Structure

Since we are concerned here with static and undrainable ponds, a feeder stream running directly into the pond should be avoided. The feeder stream must therefore be diverted along the side of the pond and from a suitable point water is channeled to the pond when required.

An inlet structure should be provided through which water can be let into the pond. A proper inlet enables the quantity of water flowing into the pond, to be regulated, preventing the entry of undesirable fish and other aquatic animals and the escape of stocked fish.

For small ponds the best inlet structure is a galvanized iron pipe of about 10 cm diameter with a control tap and a screen basket The downstream end of the pipe should be 30–40 cm above the water level. A sluice is also suitable for this purpose, especially for larger ponds. A screen is also fixed to check the entry of undesirable fishes and other animals To avoid scouring when the pond is being filled, a concrete apron can be built at the sluice, or more cheaply, a layer of gravel laid down. Similarly, if water is let in with a pipe there should be a gravel bed laid down where the water stream falls into the pond. If gravity feed is not possible, water must be pumped from the supply source into the channel leading to the pond or even directly into the pond; but, in that case, the intake should be securely wrapped by a firm net to prevent undesirable fish and other animals from entering into the pond along with the water.

MAINTENANCE

Proper maintenance of the pond and pond structure is most essential. Most of the earthen structures, especially the dykes, are susceptible to weathering action and hence they need periodical checks. Attending to minor damages regularly avoids the chances of more costly repairs later.

The grass turfing needs special attention. Proper and timely mowing prevents the formation of weedy growth and tends to develop a root system more resistant to runoff. Erosion from the top during heavy rains causes grooving out of small channels and it is an indication that the top has not been properly consolidated.The area should be levelled with more soil and thoroughly rammed and then grass should be planted to bind it.

Side erosion at the dyke bottom may be due to a number of reasons. The worst damage is done by common carp.Erosion due to frequent wave action, particularly if the grass at the edge has been grazed by grass carp, can cause undercutting of banks and subsequent collapse of dykes. Some methods used to provide protection against such erosion are earth berms, stone or brick pitching, stakes/bamboo piling.

Surface washings and organic additions cause siltation which reduces the pond depth and pond fertility. The undrainable ponds should therefore be dewatered in the summer months at the interval of 5–7 years.

FISH SPECIES CULTURE IN PONDS

Although a large number of fish species grow successfully in ponds, only a restricted number of species are usually cultivated on commercial scale. Reasons for this restricted choice is obvious.Commercial pond culture basically aims at achieving maximum possible rate of fish production and profit through optimum utilization of the natural food and the supplementary feed which drastically limits the choice of fish species for pond cultivation.

SELECTION OFSUITABLE FISH SPECIES

- Adaptability to undrainable pond environment
- Faster growth rate
- Efficient utilizers of natural food resources of the pond
- Efficient converter of artificial feed
- Hardy and not easily susceptible to disease
- Easy to breed and rear the seed
- Prolonged breeding period or multiple breeding frequency
- Non-predaceous, planktophagous and preferably herbivorous and detritus feeder
- Compatability with other cultivable species of fish
- Palatable with high nutritive value
- High market demand and high price.

To find all these qualities in one fish species would be very unlikely. Therefore, the species having maximum required traits are considered to be desirable for cultivation in undrainable ponds. Carps fit well to these criteria and hence the most widely cultivated food fishes in South Asia are the quick-growing, non-predatory carps. There are two major systems of carp culture in Asia: the Chinese polyculture system where Chinese carps are cultured together and the Indian composite fish culture system where the Indian major carps and Chinese carps are combined.

Catla

Catla is the fastest growing Indian major carp species and widely distributed throughout India, Nepal, Pakistan, Burma and Bangladesh. It inhabits the surface layer of water and feeds upon plankton. Adult stages are predominantly zooplankton feeder, occasionally taking in decaying macrovegetation, phytoplankton and smaller molluscs. It attains maturity in the second year of life and carry over 70 000 eggs per kg body weight.

It naturally breeds in rivers during monsoon season and under control conditions in bundhs as well. It does not breed in ponds.However, it responds well to hypophysation techniques. Seeds are easily reared in undrainable ponds of relatively smaller size. Under composite fish culture in ponds it usually grows to over 1 kg in one year.

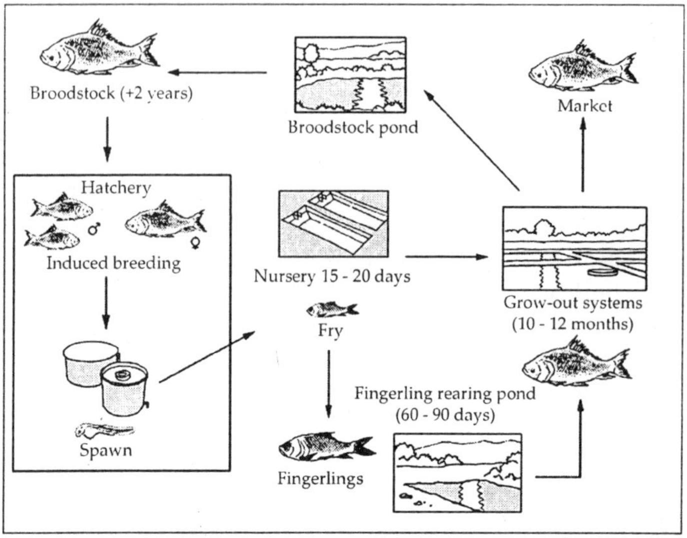

Fig. Production Cycle of Catla

Rohu

Rohu is the natural inhabitant of river systems of India, Nepal, Pakistan, Bangladesh and Burma. In recent years it has been transplanted to many countries of the world including Sri Lanka, Mauritius, USSR, Japan, Philippines, Laos, Malaysia and Thailand.

Normally it occupies the column region of the aquatic ecosystem and feeds mostly on vegetable matter including higher plants, detritus, etc. Like catla it naturally breeds in rivers and under special conditions in bundhs.Except by hypophysation to which it responds quickly, it never breeds in ponds. It attains sexual maturity during the second year.

However, certain percentages of pond-reared specimens mature within one year. Fecundity varies from 226 000 to about 2 800 000 depending upon the size. Rohu spawns during the monsoon.Seeds collected from rivers or produced by bundh breeding or induced breeding are reared with ease in seasonal or perennial undrainable ponds. Under pond culture conditions it grows upto 900 g within one year.

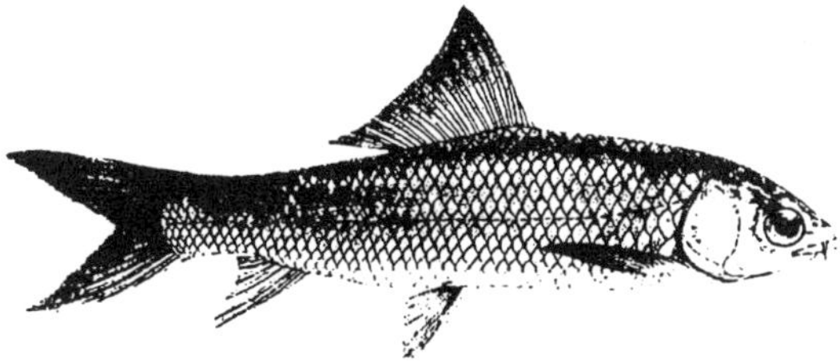

Fig. Rohu

Mrigal

Mrigal inhabits all the major river systems of India, Pakistan, Bangladesh and Burma. The adult fish feeds upon filamentous green algae, diatoms, pieces of higher plants, decayed vegetable, mud and detritus. It is basically a bottom feeder and hence suitable for cultivation with column and surface feeder carps in ponds. Mrigal usually attains maturity within 1 or 2 years depending upon the agroclimatic conditions of the location.

Fecundity ranges between 124 000 to over 1 900 000 depending upon size. Spawning season is linked with the onset and duration of the southwest monsoon. It does not breed in ponds, but can be easily bred in bundhs and by hypophysation. It is now being induced to breed twice within the same spawning season. Rearing of seed is usually undertaken in seasonal or perennial undrainable ponds. Under pond culture conditions it grows to over 1 kg in one year.

Silver Carp

Silver carp is basically inhabitant of major river systems of South and Central China and in the Amur Basin of USSR from where it has been transplanted throughout the Indo-Pacific region including India. It is a surface dweller feeding mainly upon zooplankton during its early stages and gradually becomes predominantly a phytoplankton feeder.

Its relatively longer branchiospines provide a fine filter

capable of retaining planktonic organisms. It readily accepts supplementary feed like oil cakes and rice bran mixture in pond culture systems. It does not breed in pond condition. However, through the technique of hypophysation they are induced to breed in ponds during the monsoon season. Fecundity varies greatly with the size and agroclimatic condition. A fecundity range of 145 000 to 2 044 000 has been found from silver carp.

It takes about 2–6 years to mature in China, whereas in India it matures very early, within 2 years. Males mature earlier than the females. In composite fish culture ponds it usually attains over 1.5 kg within one year of rearing. Seed rearing is done in smaller seasonal or perennial undrainable ponds with a high rate of survival.

Grass Carp

Grass carp is a native of the river systems of South-Central and North China and the Amur river of USSR. Its suitability in aquaculture and biological control of aquatic weed infestation has resulted in wide-scale transplantation throughout the world.

In early life it feeds on planktonic organisms and gradually switches over to macrophytes. They are voracious eaters and show distinct preference for vegetable food materials such as grass, leaves, weeds, etc.

However, they also accept supplementary artificial feed materials. Usually only a portion of ingested food is digested and the rest is voided in semidigested or undigested form which, in turn, becomes choice food for the bottom dweller common carp In China it takes about 3–4 years to achieve maturity whereas in India it usually takes 2 years.

The total number of eggs range between 308 800 and 618 100 from the fishes weighing between 4.7 kg to 7.0 kg. The fish does not breed under pond condition and hence seed production is achieved through hypophysation. Growth mainly depends on the rate of feeding. Under optimum feeding rate it can grow over 5 kg in one year Usually it grows to over 1.5 kg in composite fish culture ponds.

Common Carp

Originally a native of temperate region of Asia, especially China, the common carp is now the most domesticated and cultivated carp species throughout the world. It is an omnivorous bottom dweller subsisting mainly on benthic fauna and decaying vegetable matter. It frequently burrows the pond bottom in search of food.

This habit of digging the pond bottom helps in maintaining the productivity of undrainable ponds and hence culture of common carp with other carp species is of great advantage. Moreover, it also feeds directly on the undigested excreta of grass carp.

Growth mainly depends upon the bottom biota, stocking density and the rate of supplementary feed.In composite fish culture ponds it grows to about 1 kg within one year. In a tropical climate it spawns throughout the year in the pond environment with two peak periods, one from January to March and the other during July and August. Eggs are small and adhesive in nature. In tropical conditions it attains maturity within 12 months.

Index